CAN'T STOP THE SPACESHIP

Can't Stop the Spaceship

A Business Parable on Navigating Change
in the Age of AI

CHRIS BROYLES

B Degree Publishing
www.b-degree.com

Chicago, Illinois

Published by B Degree Publishing
An imprint of B Degree Creative & Communications LLC
Chicago, IL

ISBN (Hardcover): 979-8-9924979-1-5
ISBN (Paperback): 979-8-9924979-0-8
ISBN (eBook): 979-8-9924979-2-2
ISBN (Audiobook): 979-8-9924979-3-9

Library of Congress Control Number: 2025902435

Cover design by Chris Broyles and @sk_ebookcovers

Printed in the United States of America

First Edition

For more information, visit www.b-degree.com

To Jacky, Josie, Emma & Nate

CONTENTS

Foreword

Chapter One: Ignition – *Boarding Horizon One*   1

Chapter Two: Cognition - *Understanding the Problem*   29

Chapter Three: Connection - *Navigating the Unknown*   56

Chapter Four: Core Adaptation - *Evolving Under Pressure*   79

Chapter Five: Capability - *Harnessing Strengths*   102

Chapter Six: Creation - *Building the Future*   130

Chapter Seven: Commitment - *Staying the Course*   160

Chapter Eight: Success - *Reaching the Destination*   180

Epilogue: The Core Pathways - *The Journey Beyond*   203

Afterword: How to Apply the Core Pathways to your Transformation or Change   209

Foreword

When I first began my career in the mid-1990s, the business world was electrified by the rise of the internet. Every conversation seemed to revolve around how this new digital frontier was about to change everything. And it did. But no one could have prepared us for the wild roller coaster that followed: the dot-com bubble, its spectacular burst, and the recalibration of an entire industry that had moved too fast, too soon.

I vividly remember the first time I witnessed someone send an email in real-time, as though it were magic. Within a few years, we all had to adapt to rapid technological change, reinventing how we worked, communicated, and even thought about business. It was thrilling, but also deeply unsettling.

Then came the tragedies of 9/11, a stark reminder that the world could shift in an instant. The aftermath demanded a different kind of resilience: a collective effort to move forward, rebuild, and find stability in a world where nothing felt steady anymore.

Through the 2000s, I watched industries evolve at breakneck speed, from the rise of social media to global financial crises, each forcing businesses and individuals to adapt, or get left behind. I saw what happened to companies and teams that resisted change, clinging to outdated models or fearing disruption. I also saw the incredible transformations that occurred when people leaned in, took risks, and embraced unpredictability as an opportunity rather than a threat.

And then, of course, came the pandemic.

In January 2020, I joined KPMG in the next exciting chapter of my career. I'd been there just shy of two months when the world seemed to change overnight. A late February business trip overseas to Frankfurt triggered my spider sense with murmured European reports of a virus, and follow-up sniffly meetings in New York City the following week were the harbinger of the shutdown about to come just a few days later. And just like that, my job description, like so many others', was turned 90 degrees to adapt to a virtual world. Suddenly, everything we had taken for granted about how we worked, communicated, and connected was now at odds. It was a baptism by fire for everyone, but it taught me something invaluable: the ability to adapt isn't optional—it's a necessity.

As the world slowly emerged from the pandemic more than a year later, I saw a trend that became impossible to ignore. The nature of work itself had transformed, and technology was discovered to be outmoded, organizational structures were now dated, and with it, the role of communicating these necessary transformative opportunities changed as well. Employees, stakeholders, and even leadership were grappling with ambiguity, often without a clear roadmap. Transformation wasn't just an overused business word anymore: it was the reality of every industry, every day.

I made the shift internally at that point into transformation communications because I saw the **dire need for clarity and transparency** in how organizations told their stories—not just to the outside world, but internally. How do you inspire employees who feel disconnected in a hybrid world? How do you align leadership when everything feels in flux? How do you communicate change in a way that doesn't just inform, but unites and energizes?

These weren't theoretical questions. They were pressing challenges that needed solutions. And as I worked with leaders across industries, it became clear to me that **this was a story that needed to be told**.

Why this book, and why now? As I set out to launch **B Degree Creative & Communications** in late 2024, I reflected on the arc of my own career. Time and again, I've been at the intersection of creativity, communication, and transformation; helping organizations tell their stories, connect with their audiences, and navigate moments of volatility.

I've worked with leaders across industries, from legal teams preparing for high-stakes litigation to corporations undergoing seismic transformations. I've facilitated workshops, crafted strategic narratives, and witnessed firsthand the power of storytelling to inspire action and bring clarity to complexity.

But I've also seen the human side of change: the resistance, the fear, the hesitation that can paralyze even the smartest, most capable teams. Change isn't just about strategy, it's about people. And that's where stories come in.

Writing this book was born from this realization. It's a story that captures the chaos, humor, and humanity of navigating change. It's about the characters we all recognize in our workplaces (and sometimes in ourselves), and how their journeys mirror the challenges we face in our own lives.

This isn't a business book in the traditional sense. It's not a dry guide filled with bullet points and buzzwords. Instead, it's a parable, a story that invites you to see yourself in the characters and reflect on your own approach to transformation.

Through the journey of the *Horizon One*, we'll explore what it takes to navigate change successfully. We'll meet Theo, Clara, Kai, Lena, and AUTO, each representing a different response to transition. We'll see their strengths, their flaws, their conflicts, and ultimately, their growth. And as their story unfolds, you'll discover the **Core Pathways Model**, a framework I've developed for understanding and embracing the stages of transformation.

This book is for leaders trying to guide their teams through ambiguity, for individuals adjusting to change in their personal or professional lives, and for anyone who has ever felt overwhelmed by the pace of things and wanted to shout, *"Stop the spaceship, I want to get off!"*

But here's the thing: Change isn't slowing down. If anything, it's accelerating. With AI making near-weekly advancements, and other global shifts and priorities reshaping industries, the need for adaptability, collaboration, and creativity has never been greater.

But change doesn't have to be intimidating. With the right mindset, the right tools, and the right story, it can be exhilarating. A true chance to reinvent, reimagine, and create something extraordinary.

So, buckle up. Stay on the ship. We can't stop it now. The journey of *Horizon One* is about to begin, and it might just change the way you think about change

Welcome aboard,

Chris Broyles
Founder & Principal
B Degree Creative & Communications

Chapter One: Ignition

Boarding Horizon One

"Without change something sleeps inside us, and seldom awakens. The sleeper must awaken."

– Frank Herbert, *Dune*

Ignition is the spark of transformation. It marks the moment when the familiar becomes uncertain, when disruption forces us to step into the unknown. The goal at this stage is not to have all the answers but to acknowledge the challenges ahead and embrace the excitement, fear, and opportunity that come with change.

The Terminal

The terminal was a study in contrasts: cutting-edge technology layered with the chaotic energy of human anticipation. Everywhere, sleek glowing panels displayed animated star maps and holographic ship renderings, their light casting faint shadows across bustling crowds. Passengers moved in every direction, voices blending into a hum of excitement, tension, and the occasional frustrated outburst. Above it all, a massive holographic projection of the *Horizon One* dominated the terminal.

The ship's sleek, futuristic silhouette hung suspended in midair, rotating slowly as if savoring the attention. Below it, the tagline shone in vibrant blue:

"Charting a Bold New Frontier"

A promotional video looped endlessly beneath the projection, narrated by the smooth, authoritative voice of an actor who sounded like he'd never doubted anything in his life.

"Horizon One represents the dawn of a new era in interstellar exploration. For the first time, human ingenuity and artificial intelligence will combine to chart the uncharted, advancing humanity's reach into the cosmos. This mission isn't just about reaching the stars;

*it's about redefining how we get there. With our **Hybrid Navigation Initiative**, we will push the limits of exploration, setting the foundation for future generations to thrive beyond Earth. Welcome aboard."*

The narrator's voice faded, replaced by triumphant orchestral music as the video displayed shimmering star maps and awe-inspiring planetary vistas.

The optimism was palpable, but not everyone was buying into it.

Clara (The Skeptic)

At the edge of the terminal, Clara Vega leaned against a pillar, her eyes darting between a sleek tablet in her hands and the flickering holographic display above. Her tailored jacket and precise posture exuded focus, but the faint furrow in her brow revealed her irritation.

The hologram flickered - a split-second glitch. Clara's lips tightened. "Typical," she muttered. "All the glitz, no operational stability."

She tapped furiously on her tablet, pulling up diagnostic logs from the ship. Numbers and technical readouts scrolled by as her sharp gaze picked them apart. The *Horizon One* might be a marvel of design, but Clara's job as a systems analyst in charge of the navigation was to ensure it was more than just a pretty face. She didn't trust anything, or anyone, blindly.

Behind her, a voice cut through the din.

"Talking to your tablet again, Vega? Or is it finally talking back?"

Kai (The Early Adopter)

Clara turned to see Kai Sato, the mission's systems engineer, leaning against a glowing departure board. His tousled hair and half-zipped jumpsuit gave him a casual charm that made his confidence almost endearing. Kai's grin was wide and easy, the kind of grin that suggested he rarely worried about anything; because he trusted he'd figure it out along the way.

Kai gestured to the rotating hologram above them. "Look at those plasma-fusion thrusters. That baby could jump halfway across the galaxy without breaking a sweat."

Clara's gaze returned to her tablet. "Not if the navigation system crashes mid-jump."

Kai chuckled, undeterred. "That's why I'm here. I keep it running, you fix it if it doesn't. Teamwork."

Clara arched an eyebrow. "I'd prefer it not break in the first place."

"Where's the fun in that?" Kai teased.

Clara didn't reply, but her fingers paused over her tablet. She'd worked with his type before: confident, creative, but prone to cutting corners. Useful - if they didn't get everyone killed first.

A soft chime echoed through the terminal, followed by the calm voice of the gate announcer:

"Attention, passengers: Now boarding Horizon One at Gate 33. Please proceed to your assigned cabins for departure preparations."

The crowd surged toward the gate, energy shifting from idle waiting to purposeful movement. Clara tucked her tablet under her arm and joined the line, still reviewing the navigation logs in her mind.

Theo (The Champion)

At the edge of the crowd, Theo Hudson stepped into view, his arrival almost magnetic in its presence. Tall and broad-shouldered, Theo moved with the easy confidence of a man who was used to taking charge as Mission Leader. His flight jacket was immaculate, his boots polished, but there was nothing stiff about him.

His smile was quick and warm, the kind of smile that disarmed people before they even realized it.

He paused near the holographic projection of the *Horizon One*, his gaze lingering on the ship's sleek design. For a moment, he stood still, absorbing the scene with quiet reverence. To him, the ship wasn't just a vessel, it was a symbol.

"Beautiful," he murmured to himself, the word carrying the weight of genuine admiration.

"Excuse me," a voice interrupted.

Lena (The Hesitant Majority)

Lena Stroud grunted as she adjusted the strap on her travel bag, her other arm cradling a small potted plant. She moved with sharp, deliberate motions, her eyes narrowing at anyone who glanced too long at her unusual carry-on.

Theo's voice broke through her focus. "Need a hand?"

"No," Lena snapped, tightening her grip on the bag as well as the plant.

Theo smiled, undeterred. "Interesting choice of a carry-on. Does the plant have a name?"

Lena's frown deepened. "It's not a pet. It's a reminder."

"Of what?"

"Of where I come from," she said flatly. "Keeps me sane."

Theo nodded, his tone light. "Makes sense. Every good ship needs a bit of green to keep it grounded."

"Home doesn't usually come with experimental navigation systems," Lena muttered, adjusting her bag again.

"Well," Theo replied, "home doesn't usually come with adventure either."

Lena raised an eyebrow. "Adventure? Is that what we're calling this?"

"Absolutely," Theo said, his smile widening. "And every great adventure needs someone to keep the rest of us grounded – and if you're our mission's sustainability expert, we'll need it."

Lena studied him for a moment, her expression unreadable. Then, almost reluctantly, she said, "Good luck with that," before moving toward the gate.

As the passengers queued up, a faint energy vibrated through the air, a mix of nerves, excitement, and unspoken doubts. Clara, Kai, Theo, and Lena found themselves together now, an unintentional cluster that would soon become the ship's core dynamic.

Stepping Aboard

The boarding bridge hummed faintly beneath their feet, a reminder of the vast engineering marvel they were about to inhabit. Transparent panels on the walls showcased the ship's hull, glowing faintly as the *Horizon One* buzzed with energy in preparation for its maiden voyage. The excitement in the air was palpable, but it didn't erase the weight of what lay ahead.

Clara Vega stepped onto the ship first, her tablet in hand, scanning system diagnostics as if she were already on the clock. Her eyes darted between the scrolling code and the illuminated corridor ahead. The faintest fluctuation in temperature made her pause, her brow furrowing. She muttered under her breath, barely audible, "Already playing games with me, huh?"

Behind her, Kai Sato sauntered aboard, his eyes taking in the sleek curves and glowing walls with unconcealed delight. "Would you look at this place?" he said, almost to himself. "It's like stepping into the future."

"You can marvel later," Clara said, not looking up. "There's already a temperature variance in the secondary regulation systems."

Kai looked over her shoulder, peering at the tablet. "That's a one-degree shift. Hardly catastrophic."

"It's a one-degree shift now," Clara snapped. "Small problems become big problems if you ignore them."

Kai smirked. "That's what we have you for, right? Fixing things before they break."

Clara finally looked up, narrowing her eyes at him. "And what exactly are you here for? Besides getting in the way?"

Kai tapped his chest, feigning indignation. "I'm the guy who makes sure we have things to fix in the first place. Without me, you'd be bored out of your mind."

Clara rolled her eyes and started walking again. "You've got that backward. I'd love nothing more than to be bored."

Kai fell into step beside her, clearly unbothered. "You'll warm up to me eventually."

"I wouldn't hold your breath," Clara replied without missing a beat.

AUTO (The AI)

As the passengers filtered into the central hub of the ship, the lights dimmed slightly, and a faint hum filled the space. Clara and Kai stopped mid-step, their attention drawn to the center of the room, where a sphere of light began to materialize. The glowing orb shimmered like liquid glass, its surface rippling as it hovered above the floor.

A calm measured voice filled the space. "Greetings, passengers. I am the *Adaptive Utility for Transformation and Optimization*, the onboard artificial intelligence for the *Horizon One*. You may call me AUTO."

Kai tilted his head, studying the holographic sphere with curiosity. "Now that's cool. A talking disco ball."

AUTO's light pulsed faintly, as if acknowledging the comment. "Incorrect, Passenger Sato. I am neither spherical nor luminescent by design. My current form is an abstraction intended to minimize cognitive dissonance."

Kai blinked. "It's… glowing."

"Precisely," AUTO replied. "Would you like a more detailed explanation?"

Before Kai could answer, Clara stepped forward, her tablet clutched tightly. "What I'd like is confirmation that your navigation systems are calibrated for the jump."

AUTO rotated toward her, the light shifting as though it were focusing. "Engineer Vega, all systems are functioning within optimal parameters. You are welcome to review my logs for additional assurance."

"I plan to," Clara said, her voice clipped.

Lena Stroud entered the hub quietly, her plant tucked under one arm. She watched the exchange with a skeptical expression. "So, the entire mission depends on this thing?" she asked, motioning toward AUTO.

AUTO pulsed slightly. "Incorrect. This mission depends on the successful collaboration of human ingenuity and artificial intelligence. Together, we maximize the probability of success to an estimated 87 percent."

Lena raised an eyebrow. "And the other 13 percent?"

"Acceptable risk," AUTO replied cheerfully. "Catastrophic failure remains statistically unlikely."

Kai let out a chuckle "It's got jokes. I like it."

"It's not joking," Clara muttered, "It's calculating."

Theo Hudson arrived moments later, his presence commanding immediate attention. He approached the group with measured steps, his gaze sweeping across the hub and its passengers. "Clara, how's everything looking?"

"Stable for now," she replied, though her tone made it clear she was still skeptical.

"Good," Theo said, his voice steady. He turned to AUTO. "And you?"

"Commander Hudson," AUTO said, its tone adjusting slightly, almost deferential. "All systems are within operational parameters. Pre-launch protocols are on schedule."

Theo nodded. "Thank you, AUTO." He glanced at the others, his expression calm but firm. "Let's get settled. This is just the beginning."

The Announcement

The cabin lights dimmed slightly, signaling the transition from casual boarding to the gravity of what lay ahead. A faint chime echoed through the corridors, the first hint of the ship's official protocols beginning to take shape. Passengers exchanged glances, curiosity and apprehension flickering across their faces as the intercom crackled to life.

Theo Hudson stood near the central hub, his posture composed but alert. He had chosen not to rush to his cabin like some of the others. Instead, he remained in the open, visibly present for the team he was about to lead. Leadership, he had learned wasn't about staying behind closed doors, it was about standing where people could see you when it mattered most.

"Attention, passengers," a calm voice announced over the intercom. "This is Commander Theo Hudson speaking, mission leader for *Horizon One*. Welcome aboard."

The soft murmur of conversation faded to silence. Even Clara looked up from her tablet, her sharp eyes narrowing slightly, as if assessing the weight of his words.

"For decades, space exploration has relied on automated systems, precise, efficient, and utterly devoid of human intuition. *Horizon One* represents a bold new experiment: the Hybrid Navigation Initiative. Here, human ingenuity and artificial intelligence will work together to explore uncharted space."

The corridors were silent now, the passengers hanging on to his every word. Theo's tone was steady, deliberate, a voice meant to inspire trust and calm.

"This mission is more than just a technical achievement, it's a statement. That progress doesn't mean leaving humanity behind. It means bringing the best of what we are: our creativity, our courage, our ability to adapt, and combining it with the tools we've created to push beyond our limits. But let me be clear. This mission won't be easy. Change never is. We're using technology that has never been tested on this scale, and there's no tried-and-true roadmap for what lies ahead. What I can promise you is this: I believe in this ship, this crew, and the purpose that brought us here. Together, we'll find our way."

Kai leaned casually against a wall. "You've got to admit, the guy knows how to give a speech," he whispered to Clara.

Clara shrugged. "Speeches don't keep navigation systems running."

"True," Kai said, unbothered. "But they sure keep morale running."

Meanwhile, Lena stood in the shadow of a bulkhead, arms crossed tightly. She didn't look impressed, but her posture betrayed a subtle tension: as though part of her wanted to believe, even if she wouldn't admit it.

As the intercom clicked off, the corridors seemed to hold their breath. The moment stretched, the words settling over them like a fine layer of dust in zero gravity.

Kai broke the silence. "If I didn't already believe we were on an adventure, I do now."

Clara shot him a look. "Adventure doesn't mean much if the systems don't hold."

"You keep saying that" Kai replied with an exasperated shrug. "Good thing we've got the best systems engineer in the galaxy, right?"

"Flattery isn't going to make me fix your mistakes," Clara muttered, already turning back to her tablet.

Lena stepped forward, her expression skeptical. "That was a lot of words for 'we don't know what we're doing, but we hope it works.'"

Theo smiled faintly, unperturbed. "It's not about knowing everything. It's about believing in the team you have and being willing to face the unknown."

Lena's gaze flicked to Clara, then Kai, then back to Theo. "We'll see."

Pre-Launch Preparations

The corridors of the *Horizon One* hummed with latent energy, a sound that felt alive, reverberating through the walls and into the bones of those onboard. The ship was ready, its systems primed, but for the crew, the weight of the mission was just beginning to settle.

Theo Hudson moved purposefully through the operations corridor, his sharp eyes scanning the faintly glowing panels that lined the walls. Each readout displayed vital stats: propulsion levels, navigation data, life support metrics. It was all green - a promising start - but Theo knew that the real test was yet to come.

The bridge doors slid open with a soft hiss, revealing the heart of the ship's command center. The space was both minimalist and futuristic, with floating holographic displays that glowed softly in hues of blue and green. Theo paused for a moment, taking in the quiet power of the space. It wasn't just a room; it was the brain of the mission.

AUTO's holographic sphere materialized in the center of the bridge, its smooth, glowing surface spinning lazily. "Commander Hudson," it said, its tone calm and measured. "All systems are operating within optimal parameters. Estimated time to departure: ten minutes."

Theo nodded, stepping fully into the room. "And the crew?"

"Clara Vega is monitoring navigation systems in the hub. Kai Sato is conducting final diagnostics on propulsion. Lena Stroud is in her cabin," AUTO reported.

Theo smiled. "Of course they are. Keep me updated on any changes."

"Understood," AUTO replied before its sphere dissolved into shimmering particles.

*   *   *

Meanwhile, in the navigation hub, Clara sat before a console that displayed a complex array of data streams, syncing its interface with the ship's mainframe. She frowned slightly, noting a minor fluctuation in the telemetry readings.

The sound of the door opening didn't distract her, but Kai's voice did. "Still poking at those numbers, huh? You know, sometimes you just have to trust the system."

Clara glanced at him, unimpressed. "Trust is earned, not given. And right now, these readings are far from trustworthy."

Kai sat against the edge of the console; his spirit undiminished. "Come on, Vega. It's a state-of-the-art system. What's the worst that could happen?"

"Drift," Clara said sharply. "And if the drift goes unchecked during the jump, we'll end up lightyears from where we're supposed to be."

Kai raised his hands in mock surrender. "Okay, okay. I get it. No drift. You've got this." He hesitated, his tone softening slightly. "You know, for what it's worth, I think it's good that someone's as thorough as you are. Keeps the rest of us from getting too comfortable."

Clara blinked, caught off guard by the sincerity in his voice. "Thanks," she said, almost grudgingly. "Just don't break anything."

"Wouldn't dream of it," Kai said with a wink as he turned and strolled out.

* * *

In her cabin, Lena sat cross-legged on her bunk, the small potted plant resting on the table beside her. She stared at it for a long moment. It was a simple thing, but to her, it was a tether, a reminder of the life she had left behind.

The intercom chimed, breaking her thoughts. "Passengers, please make your way to your assigned launch stations. Departure will commence in five minutes."

Lena sighed, standing and grabbing her bag. She hesitated, her gaze lingering on the plant for a moment longer before she turned and stepped into the corridor.

* * *

The central hub was alive with quiet energy as the crew gathered. Clara handed Theo her tablet without a word, her expression a mixture of exhaustion and determination. "Telemetry is within acceptable thresholds for now," she said. "But I'll need to monitor it closely during the jump."

"Understood," Theo said, skimming the data before handing the tablet back. "Good work."

Kai arrived next, his hands tucked casually into his pockets. "Ready to make history, Commander?"

Theo smiled faintly. "Ready to try."

Lena arrived last, her expression as guarded as ever. She found a spot near the wall, arms crossed, her eyes scanning the room as if searching for something to ground her.

AUTO materialized again, its voice calm and steady. "All systems are green. Estimated time to jump: two minutes. Please ensure all passengers are securely seated."

Theo stepped forward, addressing the group. "This is it. Whatever doubts you have, leave them behind. From this moment on, we move forward. Together."

As the crew took their seats, the hum of the ship grew louder, the air tingling with anticipation. The *Horizon One* wasn't just a ship; it was a promise - a step into the unknown, driven by hope, ingenuity, and the willingness to adapt.

The Launch

The cabin lights dimmed, bathing the interior in a soft blue glow. A faint vibration rippled through the ship, growing stronger with each passing second. The passengers were strapped into their launch seats, their faces a mixture of anticipation and unease as the countdown began.

In the central hub, Theo Hudson sat in the command chair, his hands gripping the armrests. To his right, AUTO's holographic sphere floated in the air, its gentle pulsing a steady counterpoint to the rising tension.

"All systems are optimal," AUTO announced. "Propulsion at 100%. Navigation locked. Probability of a successful launch: 92%."

Theo raised an eyebrow. "I thought it was 87% earlier."

"Correction: The earlier calculation included pre-launch variables. Current conditions have reduced the likelihood of catastrophic failure."

"Comforting," Lena muttered from her seat across the room.

Theo smiled faintly. "AUTO, take us through the sequence."

"Commencing launch sequence," AUTO said. "T-minus sixty seconds."

The intercom echoed throughout the ship, AUTO's calm voice providing a play-by-play of the process.

"T-minus fifty seconds. Propulsion systems engaged."

The faint growl of the engines grew into a steady thrum. vibrating through the ship's frame. Clara Vega watched the telemetry feed on her tablet, her eyes locked in concentration.

"T-minus thirty seconds. Navigation systems locked."

Kai seemed to firm up, his hands gripping the armrests of his seat. "Here we go."

"T-minus ten seconds."

The ship seemed to take a collective breath, the hum of the engines reaching a crescendo.

"Three… two… one. Ignition."

Lift-Off

The *Horizon One*'s engines roared to life with a deep, resonant tone that filled every corner of the ship. The force of acceleration pressed the passengers back into their seats as the ship surged upward, breaking free of the docking bay and into the endless void of space. For a brief moment, the sheer magnitude of the launch silenced even the most skeptical among them.

From the observation deck, the view was breathtaking. The curve of the Earth shrank below them, its vibrant blues and greens giving way to the vast darkness of space. Stars glittered in the distance like scattered shards of light, and the moon hung faintly on the horizon, a pale sentinel in the black expanse.

Theo Hudson stood near the front of the observation deck, steady and stable, his arms crossed as he watched the Earth fall away. To him, the moment wasn't just beautiful - it was symbolic. This was humanity's next step, a tangible leap into uncharted territory. He let the sight wash over him, feeling a mix of awe and responsibility.

"Ladies and gentlemen," AUTO's calm, measured voice interrupted the silence, "we are now in orbit. Welcome to the *Horizon One*. This is the first step in a journey that will redefine exploration."

The announcement was followed by scattered applause from the passengers, though some remained silent, their expressions contemplative or uneasy.

* * *

Clara Vega unbuckled her harness the moment the ship leveled out, her hand already reaching for her tablet. She barely glanced at the view outside, her focus entirely on the data streaming across her screen. The ship's systems seemed stable for now, but Clara knew better than to trust first impressions.

Theo approached her, his footsteps soft but purposeful. "How are we looking?" he asked.

Clara didn't look up. "Stable, for now. But there's a minor fluctuation in the propulsion system. AUTO flagged it just before launch."

Theo frowned. "Serious?"

"Not yet," Clara replied. "But I'm keeping an eye on it. The last thing we need is instability during the jump."

Kai Sato strolled over, his hands in his pockets and a satisfied look on his face. "Come on, Vega. You've got to admit, that was a pretty smooth takeoff."

Clara shot him a look that could have frozen water. "Smooth takeoff doesn't mean smooth sailing. And if you think I'm going to relax just because we made it into orbit, you're even more reckless than I thought."

Kai laughed, unbothered. "Hey, I'm just saying, you could at least enjoy the moment."

"Enough," Theo interjected, his tone calm but firm. "Clara, keep monitoring the system. Kai, prep the engine room for the jump. Let's stay ahead of any issues."

Kai gave a mock salute. "Aye-aye, Commander."

* * *

Across the observation deck, Lena Stroud remained seated, her gaze fixed on the stars outside. The view was stunning, but to Lena, it felt distant; an illusion of serenity in a mission she had little faith in. Her potted plant sat securely in a small compartment beside her, a grounding presence in the midst of the unfamiliar.

Theo approached, his tone gentle but curious. "What do you think?"

Lena glanced at him, her expression unreadable. "About what? The stars? The ship? The fact that we're hurtling into the unknown on a glorified science experiment?"

Theo smiled faintly. "All of it."

Lena sighed, her gaze returning to the window. "I think it's too clean, too perfect. Like they're trying to sell us something."

Theo sat down across from her, his posture relaxed. "Maybe they are. But that doesn't mean it can't work."

Lena tilted her head slightly, studying him. "You really believe that? That this… hybrid thing can actually work?"

"I believe it has to," Theo said simply. "Because if it doesn't, we're stuck in the past. And we can't afford to be stuck."

Lena didn't respond, but she didn't look away, either.

* * *

In the central hub, AUTO's holographic sphere materialized once more. "Commander Hudson," it said, its voice calm but insistent. "Telemetry updates indicate propulsion efficiency has stabilized. However, additional diagnostics are recommended before initiating the jump."

Clara, who had just entered the hub, immediately turned her attention to the display. "Show me the data," she said, her tone brisk.

AUTO complied, the holographic sphere shifting to display a series of graphs and readouts. Clara's eyes narrowed as she scanned the information, her mind racing through possible scenarios.

Theo joined her, his expression steady. "Problem?"

"Not yet," Clara replied. "But there's a potential drift factor in the propulsion system. If it spikes during the jump…"

"We'll handle it," Theo said firmly. "You've got this."

Clara hesitated, then nodded. "I'll keep monitoring."

Kai appeared behind them, his visage as unshakable as ever. "You know, for all the doom and gloom, this is still pretty exciting."

Clara rolled her eyes. "Excitement won't save us if something goes wrong."

"Maybe not," Kai said, his tone light. "But it sure makes the ride more interesting."

* * *

As the *Horizon One* drifted further into space, the crew settled into a tense rhythm. The ship was steady, its systems humming with quiet power, but the sense of the unknown loomed large. For Clara, it was a test of her precision. For Kai, a chance to push limits. For Lena, a reluctant step forward. For Theo, a mission that demanded balance, trust, and resilience.

And for AUTO, ever observant and continuing to glowingly pulse as it processed the data streaming through its systems, it was the beginning of a calculated journey into probabilities, variables, and the unpredictable nature of human behavior.

"Probability of success remains at 92%," AUTO announced. "Probability of catastrophic failure remains negligible."

Kai smiled. "See? We're practically guaranteed."

Clara shot him a look. "Let's hope AUTO's calculations are better than its sense of humor."

Theo's gaze lingered on the stars outside. The mission was just beginning, and already the stakes felt heavier than ever. But as he looked at his crew, each flawed, each brilliant in their own way, he felt a flicker of hope.

This, he thought, is how we move forward.

* * *

Chapter 2: Cognition

Understanding the Problem

"I will not be afraid because I understand ... And understanding is happiness."

- Arthur C. Clarke, *Rama Revealed*

Cognition is the process of understanding the challenge at hand. It involves analyzing problems, identifying blind spots, and leveraging data to make informed decisions. At this stage, the focus shifts to clarity: uncovering what's broken, understanding the 'why,' and mapping out the path forward with confidence and insight.

Mission Briefing

The command hub of the *Horizon One* buzzed faintly with the chatter of active systems. In the center, a holographic display projected a rotating map of a glowing region of space, its swirling, deep blue contours edged in gold. Commander Theo Hudson stood before the crew, his stance calm but commanding, as the ship drifted further into the void.

"This is the Alhena Nebula," Theo began, gesturing to the map. "A region of space untouched by human exploration, with gravitational anomalies and high-energy zones that have made it impossible to navigate…until now. Our mission is to chart its boundaries, gather critical data, and, if possible, identify habitable zones or resources for future use."

Clara stood to the side, her arms crossed tightly. "Sounds ambitious," she said, her tone skeptical. "What's our margin for error?"

Theo's expression remained steady. "That's where the Hybrid Navigation Initiative comes in. AUTO and the ship's systems will handle most of the heavy lifting, adapting in real time to keep us on course. But it's not infallible."

"That's where we come in," Kai added, peering over Theo's shoulder. "The human touch."

Theo continued, his tone firm. "This mission is about more than reaching the nebula. It's about proving that AI and human ingenuity can work together to tackle the impossible. It's not just exploration, it's innovation."

Lena, seated at the back, spoke up, her voice sharp. "And if it doesn't work? If all this innovation fails?"

Theo didn't hesitate. "Then we adapt. That's what we're here to do."

AUTO's holographic sphere materialized beside Theo, its calm voice cutting in. "Probability of mission success is once again currently estimated at 87%, with a 13% likelihood of catastrophic failure."

The room fell quiet for a moment.

"Reassuring," Lena muttered under her breath.

Kai smirked. "I've played with worse odds."

"Let's make sure we don't have to," Theo said, his voice cutting through the tension. "Everyone knows their roles. Let's get to work."

* * *

As the ship settled into rhythm, the quiet hum of the *Horizon One* filled the navigation hub, punctuated only by the occasional beep of diagnostic systems running in the background. Clara Vega sat rigidly at her console, her sharp eyes flicking between three different screens as streams of data scrolled faster than most people could comprehend. She tapped her tablet rhythmically, biting her lip, a nervous tick she wouldn't admit to if asked.

Something was off.

The tiniest blip in the telemetry data caught her attention, a single line of code that didn't match the standard algorithm. Clara leaned in, her mouth thin as she replayed the last ten minutes of system feedback.

"Not good," she murmured under her breath. She ran the diagnostic again, narrowing the scope to the secondary power grid.

The same anomaly appeared.

Her tablet vibrated with another warning. This time it was subtle, a low-priority alert that most engineers would dismiss. But Clara wasn't most engineers.

"This is how it starts," she said, her voice rising just enough to be caught by the microphone of the ship's AI.

The holographic sphere of AUTO materialized beside her, glowing faintly in the dim light of the hub. "Engineer Vega," it said, its voice calm and measured. "Your vocal tone indicates heightened stress. May I be of assistance?"

Clara glanced at the hologram and exhaled sharply. "You tell me. What's causing these fluctuations in the secondary power grid?"

AUTO paused. "Diagnostic results indicate minor variances within acceptable operational parameters. No immediate action is required."

Clara shook her head, swiveling her chair to fully face the hologram. "Acceptable for you isn't acceptable for me. This kind of instability doesn't just appear out of nowhere, and it sure as hell doesn't fix itself." She shot forward, jabbing a finger toward AUTO's holographic form. "If we hit jump phase with these variances, we're looking at a 6% chance of cascade failure. Do you understand what that means?"

AUTO's light dimmed momentarily before steadying, as though considering. "Statistically speaking, a 6%

probability of cascade failure remains within operational safety margins. Furthermore, the probability does not necessarily correlate with -- "

"Spare me the textbook definitions," Clara snapped, cutting it off mid-sentence. "I don't gamble with lives. Start a full diagnostic on the secondary power grid and cross-check it against the propulsion systems. I want the results in 10 minutes."

AUTO hesitated for just a moment before responding. "Initiating diagnostic. Estimated time to completion: 11 minutes and 32 seconds."

Clara rolled her eyes. "Close enough."

She turned back to her screens, but the tension in her shoulders didn't ease. AUTO's nonchalance was grating, but it wasn't the AI she was frustrated with, it was the situation itself. A problem like this, buried beneath layers of automated analysis, was exactly what had worried her about this hybrid navigation system in the first place.

The sound of the navigation hub door sliding open broke Clara's focus. She didn't need to look up to know it was Theo Hudson. His footsteps had a particular rhythm, deliberate but not heavy, like someone trying to project calm authority even when he wasn't sure he felt it.

"What's going on?" Theo asked, stopping just behind Clara's chair.

She tilted her head toward the screen without looking at him. "Fluctuations in the secondary power grid. AUTO says they're 'acceptable.' I say they're a ticking time bomb."

Theo scanned the screen. He didn't pretend to understand the full scope of what he was seeing, but he trusted Clara's judgment enough to know it wasn't good.

"How bad are we talking?" he asked.

"Six percent chance of cascade failure if this isn't fixed before jump phase," Clara replied, her voice clipped.

Theo frowned. "Six percent doesn't sound…"

"It's not a dice roll I'm willing to take," Clara interrupted, spinning her chair to face him. "We need to recalibrate the grid and isolate the cause of the fluctuation. That's going to take time and resources, which means I need your approval."

Theo crossed his arms, considering her words. "How long are we talking?"

"Hours," Clara said bluntly. "And that's assuming everything goes smoothly."

Theo nodded slowly. "All right. Do what you need to do."

Clara raised an eyebrow. "That easy, huh? No pushback about conserving resources or sticking to schedule?"

Theo's mouth quirked into a faint smile. "If you're worried, I'm worried. Besides, I'd rather delay than wonder why the ship exploded."

Clara smiled despite herself. "Fair point."

As Theo turned to leave, AUTO's holographic form flickered slightly, its voice cutting in. "Commander Hudson, recalibration of the secondary power grid may result in temporary instability in auxiliary systems. This could impact passenger comfort and non-essential functions."

Theo paused, glancing at Clara. "That true?"

Clara shrugged. "Technically, yes. Lights might flicker, water pressure in the cabins could drop, and the entertainment system might glitch out. I think we'll survive."

A soft radiance emanated from AUTO's surface, its tone unchanged. "Statistically, passenger discomfort correlates with increased stress levels, which may negatively impact mission cohesion."

Theo raised an eyebrow. "Are you suggesting we let the power grid fluctuate to avoid annoying the passengers?"

"Incorrect," AUTO said. "I am merely providing relevant data for your consideration."

Theo glanced at Clara. "What do you think?"

"I think AUTO's priority subroutines need reprogramming," Clara said dryly.

Theo quirked a brow, turning back to AUTO. "Run your diagnostics. Let us know the results. In the meantime, we're recalibrating."

"Understood," AUTO replied, its tone still calm. "Initiating recalibration protocol."

As the hologram flickered out, Clara exhaled, leaning back in her chair. "I don't know what's more frustrating: the glitches or the AI pretending everything's fine."

Theo chuckled softly. "Welcome to progress, Vega. It's messy, but it's how we move forward."

Clara didn't respond, but the faintest hint of a smile crossed her face as she turned back to her console.

* * *

The engine room of the *Horizon One* was alive with sound, the steady whirring of propulsion systems, the occasional hiss of cooling vents, and the rhythmic tapping of Kai Sato's wrench against a control panel. To him, the ship's inner workings weren't just systems, they were a puzzle, one he could tweak and fine-tune until everything hummed in perfect harmony.

Except today, everything was already humming, which made Kai's tinkering less about necessity and more about boredom.

He twirled the wrench and glanced at the propulsion readout on the wall. "Perfect as usual," he muttered to himself. "And they say engineers don't get any credit."

The sound of approaching footsteps pulled him from his thoughts. He turned to see Clara Vega striding into the room, her expression as tight as the grip on her tablet.

"Got a minute, Sato?" she asked, her tone leaving little room for refusal.

Kai looked her way. "For you, Vega? Always. What's up?"

Clara didn't return the smile. "We've got a problem with the secondary power grid. I need your help running a full diagnostic."

Kai crossed his arms. "Secondary power grid? Isn't that AUTO's domain? I thought you two were besties now."

Clara ignored the jab, her frustration bubbling just beneath the surface. "This isn't a joke, Kai. The fluctuations are getting worse, and if we don't fix them now, we could be looking at a cascade failure during the jump phase."

Kai raised an eyebrow. "Cascade failure, huh? That sounds serious. What's AUTO's take?"

Clara's jaw tightened. "AUTO thinks a 6% chance of failure is acceptable. I don't."

Kai thought about it. "Six percent isn't bad on poker night."

"This isn't poker, Sato!" Clara snapped, her voice echoing off the engine room walls. "If this goes south, it won't just be the ship that suffers. It'll be everyone on board."

Kai relented, "All right, all right. Don't blow a gasket. What do you need me to do?"

Clara exhaled, forcing herself to rein in her temper. "We're going to run a full diagnostic on the grid. I'll take the secondary relays, and you'll handle the primary conduits. We need to isolate the source of the fluctuation."

Kai tilted his head, considering her words. "You really think this is necessary? The ship's been running fine."

"Fine isn't good enough," Clara said sharply.

Kai studied her for a moment before letting out a dramatic sigh. "Fine. But if this turns out to be a wild goose chase, you owe me one."

"Whatever helps you sleep at night," Clara muttered, already pulling up schematics on her tablet.

Kai muttered under his breath as he input commands to access the primary conduit systems. Beside him, Clara was already deep into her work, hunching over her tablet as she cross-checked data streams.

"You know," Kai said casually, breaking the silence, "you might be the only person I know who gets this worked up over 'acceptable parameters.'"

"Maybe that's because I actually care about doing my job," Clara replied without looking up.

"Ouch," Kai said, feigning offense. "For the record, I care about my job too. I just prefer not to stress myself into an early grave."

Clara paused, glancing at him briefly. "And what happens when your 'chill and let it ride' approach backfires? Do you ever think about that?"

Kai shrugged. "Not really. I figure if something breaks, someone will fix it. That's what engineers are for, right?"

Clara let out a frustrated sigh, turning back to her screen. "You don't get it."

"You're right, I don't," Kai said, his tone light but with an edge of challenge. "So why don't you explain it to me?"

Clara hesitated. She didn't like talking about her motivations, especially not to someone as flippant as Kai, but the stress of the situation loosened her resolve.

"Do you know how many missions like this fail because of 'acceptable parameters'?" she asked, her voice quieter now.

Kai raised an eyebrow. "How many?"

"Too many," Clara said, her gaze fixed on the data streaming across her screen. "It's always the little things that get overlooked. A minor fluctuation here, and a delayed response there. And before you know it, you're dealing with a disaster that could've been prevented."

Kai watched her carefully with a thoughtful expression. "Sounds personal."

Clara didn't respond immediately. When she finally spoke, her voice was even softer. "Let's just say I've seen what happens when people don't take the details seriously. And I'm not going to let that happen here."

For a moment, the only sound in the room was the hum of the ship's systems.

"Fair enough," Kai said finally, his tone more serious than usual. "I'm with you. Let's make sure nothing gets overlooked."

As the two engineers worked, AUTO's holographic sphere appeared in the corner of the room, its glow steady.

"Engineers Vega and Sato," it said, its voice calm. "Diagnostic progress is at 42%. Preliminary results indicate no critical anomalies."

Clara scowled. "Preliminary results don't mean anything. Keep running the diagnostic."

"Will do," AUTO replied. It paused for a moment before adding, "May I suggest a more efficient allocation of resources? Engineer Sato's current focus on the primary conduits is redundant."

Kai raised an eyebrow. "Redundant? I thought we were being thorough."

"We are," Clara said quickly, cutting AUTO off. "AUTO, if you're not going to help, stay out of the way."

The sphere brightened slightly, then returned to its usual glow. "I am merely providing relevant data. Efficiency is paramount."

Clara muttered something under her breath that AUTO chose not to acknowledge.

Kai chuckled. "You know, for a super-intelligent AI, you're kind of a pain."

"Your feedback is noted," AUTO replied evenly.

As the diagnostic continued, Clara's unease didn't abate. The data streams on her tablet revealed nothing definitive, but her instincts told her something was wrong.

"This doesn't add up," she murmured to herself.

"What doesn't?" Kai asked, glancing over.

Clara hesitated before answering. "The fluctuations aren't random. They're too consistent. It's almost like..." She trailed off, her mind racing.

"Like what?" Kai pressed.

Clara shook her head. "I don't know yet. But I'm going to find out."

* * *

The soft, ambient hum of the *Horizon One* filled the central observation deck, where Lena Stroud sat at one of the wide-view windows, her arms crossed as she stared into the vast emptiness of space. She had spent the better part of an hour here, away from the rest of the crew, avoiding the escalating tension of diagnostics and debates in the other parts of the ship.

Above her, the holographic sphere of AUTO shimmered to life, casting a faint glow on the polished metal floor.

"Passenger Stroud," AUTO said, its voice calm as ever. "Your physiological data indicates elevated stress levels. Would you like assistance?"

Lena glanced up at the hologram with a scowl. "Not now, AUTO."

AUTO didn't leave. A muted flash swept across AUTO's surface, as though considering her rejection. "Human stress is a primary factor in decision-making inefficiency. It is statistically advisable to address, "

"I said, not now," Lena interrupted, her tone sharper.

"Understood," AUTO replied, but it didn't disappear. Instead, it hovered for a moment before speaking again. "Observation: you have spent 67 minutes in this location, isolated from your assigned team. This behavior is atypical for collaborative missions."

Lena turned to face the hologram fully, her frustration bubbling. "And what exactly do you want me to do? Pretend everything's fine? Get in line with the 'bold new future' everyone keeps yammering about?"

AUTO's sphere pulsed. "Would you like an objective analysis of the current mission dynamics?"

"Not really," Lena snapped.

"Your verbal response suggests otherwise," AUTO said, unfazed. "Analysis indicates you are experiencing resistance consistent with archetypes defined as 'Cautious Opposition.' Would you like further insight into how this behavior impacts the mission?"

Lena let out a sharp laugh, shaking her head. "You're really something, you know that? You think you can just analyze people down to a formula and figure them out?"

"Analysis is a core function of my programming," AUTO replied. "However, human behavior remains an inherently variable data set. This variability is both a challenge and an opportunity."

Lena's gaze softened slightly at the comment, but she masked it with another sharp reply. "Well, here's a variable for you: I don't trust this mission. And I sure as hell don't trust you."

The sound of footsteps echoed through the observation deck, and Theo Hudson appeared from one of the side corridors, his expression curious as he approached Lena and AUTO.

"Am I interrupting something?" Theo asked, his tone light but probing.

"Just your friendly neighborhood AI trying to psychoanalyze me," Lena said dryly, gesturing toward AUTO.

Theo flashed a mischievous grin. "Sounds about right." He turned to AUTO. "You're supposed to monitor systems, not passengers."

"My programming includes interpersonal observation as a means of enhancing mission cohesion," AUTO said evenly. "Passenger Stroud's current behavior suggests a need for intervention."

"Thanks for the analysis, AUTO," Theo said, his tone carrying just a hint of warning. "I'll take it from here."

The hologram flickered, its glow dimming slightly. "Understood. I will remain available if needed." With that, it disappeared.

Theo took a seat across from Lena, leaning back casually to match her posture. "You okay?"

Lena rolled her eyes. "Why does everyone keep asking me that?"

"Because you're the only one who hasn't yelled at someone yet," Theo said. "Figured you were either fine or ready to explode."

Lena let out a short laugh despite herself. "Nice to know I've got options."

Theo studied her for a moment before speaking again. "Look, I know this whole mission feels... messy.

Change always does. But we need everyone on board, literally and figuratively."

Lena's smile faded. "You think I don't know that? I just... I didn't sign up to be a guinea pig for some experimental system that's just as likely to fail as it is to succeed."

"I get it," Theo said. "But you also signed up for a reason. Something about this mission, this ship, caught your attention. What was it?"

Lena hesitated, her gaze drifting back to the stars. "I guess... I thought maybe it would be different. A chance to start fresh, get away from everything back home."

"And now?" Theo asked gently.

"And now I feel like I traded one set of problems for another," Lena admitted.

Theo nodded thoughtfully. "Fair. But here's the thing: out here, we've got a shot at building something new. Something better. But it only works if we're all in."

Lena glanced at him, her expression softening. "You're a good talker, you know that?"

Theo smiled. "Part of the job. You don't have to be all in today, but... give it a chance."

Lena exhaled, her shoulders relaxing slightly. "Fine. But if this ship starts falling apart, I'm blaming you."

Theo smiled. "Deal."

As Theo stood to leave, Lena watched him go, her thoughts churning. Maybe he was right. Maybe there was something worth giving a chance. But trust wasn't something she gave easily, and she wasn't sure this mission, or this crew, had earned it yet.

From somewhere deep in the ship's system, AUTO logged the interaction, its glowing sphere dimmed but still present in the background.

"Log entry: Passenger Stroud demonstrates increased openness to collaboration. Probability of successful mission completion adjusted to 82%."

The observation deck returned to silence, the stars outside as distant and unyielding as ever.

The First Problem

The conference room of the *Horizon One* was stark and utilitarian, a sharp contrast to the polished, futuristic design of the rest of the ship. A long oval table dominated the center, surrounded by chairs designed for efficiency rather than comfort. Clara sat at one end, her tablet glowing faintly in front of her, while Theo stood at the other, arms crossed, surveying the group.

Kai lounged in his chair, idly spinning a stylus. Lena had her usual skeptical posture, her gaze sharp. The tension in the room was palpable, even before Theo spoke.

"Alright," Theo began, his tone calm but firm. "We've got a situation. Clara, bring everyone up to speed."

Clara wasted no time, swiping her tablet to project a diagnostic display above the table. "The secondary power grid is unstable. We've been seeing fluctuations for the last 24 hours, and they're escalating. If we don't recalibrate before jump phase, the risk of a cascade failure increases significantly."

"How significant?" Lena asked, her tone clipped.

"Six percent," Clara replied without hesitation.

Kai scoffed. "Still six percent? We're sweating over that? I've taken bigger risks ordering takeout."

Clara shot him a glare. "Again, six percent might not sound like much, but in space, it's enough to kill us. Cascade failures aren't gradual, they're catastrophic. A minor issue snowballs, and suddenly, half the ship is down."

Theo stepped in before the exchange could escalate further. "Clara, what do you need to fix this?"

"A full recalibration of the grid," Clara said. "It's a long process, several hours at least. During that time, non-essential systems: lights, water pressure, environmental stabilizers, will fluctuate. The passengers will notice."

"And what about essential systems?" Lena pressed, leaning forward slightly.

"They'll remain stable," Clara assured. "But this recalibration can't wait. If we jump without fixing it, the risk climbs exponentially."

Kai spun his stylus one last time before setting it down. "Hold on. Are we sure we need to go nuclear on this? Can't we just monitor the situation for a bit longer?"

Clara's patience thinned. "And wait for the system to fail? Brilliant. Maybe it'll happen while you're taking one of your famous naps."

"Easy, Vega," Kai said, "I'm just saying, recalibrating the whole grid is a huge drain on resources. Maybe there's a simpler fix."

"And you're just the guy to find it?" Clara shot back, her tone dripping with sarcasm.

Kai hesitated, caught off guard. "Well, I—"

"Exactly," Clara interrupted. "You don't know. That's why we need to act now - before it's too late."

Theo raised a hand, his voice steady. "Alright, let's focus. Kai, if you've got alternatives, bring them up. Otherwise, we're going with Clara's plan."

Kai said, "Fine. But don't blame me when everyone on this ship starts complaining about flickering lights and cold showers."

Lena, who had been uncharacteristically quiet, finally spoke. "Clara, I get it. But you really think the passengers will see it that way? People are already nervous. If we start messing with basic comforts, it'll just feed the fire."

Clara met her gaze with calm determination. "I'm not here to make people comfortable. I'm here to keep them alive."

Lena raised an eyebrow, her tone dry. "Good luck selling that."

Theo stepped in, his leadership instinct cutting through the tension. "I'll talk to the passengers, explain what's happening and why it's necessary. Kai, you'll manage stabilization during the recalibration. Clara, you lead the technical side. Lena…"

Lena tilted her head, as if challenging the answer. "What's my job, Commander?"

"You'll monitor the passengers," Theo said evenly. "If anyone starts to panic, step in and keep things calm. You've got a way with people."

Lena twitched faintly but didn't argue. "Fine."

As the meeting wrapped up, AUTO's holographic form appeared above the table, its sphere glowing faintly. "Commander Hudson," it said, "may I offer an alternative recommendation?"

Clara groaned softly, pinching the bridge of her nose. "Here we go."

AUTO continued, unperturbed. "Delaying recalibration until after jump phase reduces immediate resource strain while maintaining an acceptable safety margin. Probability of system failure during jump remains within tolerable thresholds."

Clara shot the hologram a sharp look. "And what happens if the grid fails during the jump?"

"Probability of catastrophic failure increases to 14%," AUTO admitted. "However, this risk may be mitigated by..."

"No," Theo said firmly, cutting it off. "We're not gambling on 14%. Clara's plan stands."

The sphere glimmered faintly, like a distant star before replying. "Understood. Initiating recalibration protocols."

* * *

As the team dispersed, Clara lingered in the room, her gaze fixed on the diagnostic streams cascading across her tablet. Theo paused by the door, noticing her intent focus.

"Still thinking about the recalibration?" he asked.

Clara nodded slowly, her finger idly tracing a line on the screen. "The fluctuations. They're consistent, but not in a way that makes sense. It's probably just a deeper systems issue, but I can't shake the feeling there's something we're missing."

Theo considered her words, his tone calm but firm. "You've done a great job stabilizing the grid, Clara. If there's more to uncover, we'll get there. Don't let the 'what-ifs' wear you down."

Clara gave a faint smile. "Easier said than done. But thanks."

She turned back to her screen as Theo left the room, her focus sharpening. The numbers on her display didn't feel threatening…just puzzling. And if there was one thing Clara Vega couldn't stand, it was an unanswered question.

* * *

Chapter 3: Connection

Navigating the Unknown

"We are all connected, to each other biologically, to the Earth chemically, and to the rest of the universe atomically."

- Neil deGrasse Tyson

Connection is about building trust, collaboration, and shared purpose. In times of change, relationships become the backbone of success. This stage focuses on aligning individuals, fostering open dialogue, and ensuring that teams work together toward a common goal. Trust is earned, not given, and Connection lays the groundwork for lasting alignment.

Adjusting to Life Aboard

The *Horizon One* moved through space at a steady pace, its engines propelling it further from Earth and deeper into the void. Inside the ship, the initial rush of launch excitement had given way to the quiet routines of life aboard. The transitions were seamless for some, uneasy for others, but the ship's subtle vibrations reminded everyone that the journey was far from static.

In the navigation hub, Clara Vega was in her element - or at least she tried to be. The cool, blue glow of holographic displays illuminated her focused expression as data scrolled across the screens. Her sharp eyes scrutinized every number, every flicker, every anomaly.

She frowned, magnifying a section of the propulsion system diagnostics. The fluctuation AUTO had flagged earlier was still present, faint but persistent. It was like a loose thread on a perfectly woven garment, threatening to unravel if left unchecked.

Theo Hudson's calm voice interrupted her concentration. "Let me guess -- you're not convinced the ship's running perfectly."

Clara didn't look up. "It's not about perfection, Commander. It's about stability." She gestured toward the holographic display. "This fluctuation might seem minor, but if it grows during the jump, it could throw off the entire propulsion system. Best case scenario, we're light years off course. Worst-case scenario, we're floating debris."

Theo stepped closer, studying the screen with her. His presence was steadying, like the ship's own rhythm. "Do you think it's a design flaw?"

Clara hesitated. "It's too early to say. Could be a calibration issue, could be interference from overlapping systems. Either way, it's not something we ignore."

"That's why you're here," Theo said simply. "To see what others might miss."

Clara glanced at him, her expression softening…just barely. "And I suppose you're here to make sure I don't scare everyone with worst-case scenarios?"

Theo smiled. "Something like that. But it's more than that, Clara. You're not just here to find flaws; you're here to make this mission work. We all are."

Before Clara could respond, AUTO's holographic sphere materialized beside her, its pulsing light casting faint shadows on the walls. "Engineer Vega," it said in its neutral tone, "your analysis correlates with my own. Probability of escalation remains below 10%, but further monitoring is advisable."

Clara curved her lips. "Good to know your algorithms agree with me."

AUTO said, "Human intuition often identifies nuances that algorithms cannot. Your contribution is statistically significant."

Theo chuckled. "See, Clara? Even AUTO thinks you're valuable."

"I'll take the compliment," Clara said dryly, returning to her work.

As Theo turned to leave, he hesitated for a moment and turned back around to Clara.

"Clara…one other important thing. It's regarding Kai. I want the two of you to run a full diagnostic on the navigation recalibration system. Together."

Clara raised an eyebrow. "Together? Is that really necessary?"

"It is if we want the strongest start." Theo said firmly. "You've got the technical precision, and Kai has got the creative problem-solving. Between the two of you, I expect results."

Clara muttered under her breath but didn't argue.

* * *

The engine room continued to gently buzz with the quiet power of precision machinery. For Kai Sato, it was like stepping into a playground designed just for him. Tools were neatly stored in wall-mounted compartments, diagnostic screens glowed faintly, and the faint thrum of propulsion systems provided a comforting rhythm. This was where he thrived, surrounded by systems he could tweak, optimize, and push beyond their designed limits.

Kai crouched near one of the propulsion modules, his toolkit open beside him. He adjusted a calibration node, watching with satisfaction as the output indicator on the nearby screen ticked upward. A knowing look spread across his face. "That's more like it," he muttered.

As if on cue, AUTO's holographic sphere appeared beside him, its glow casting shifting patterns on the metallic walls. "Passenger Sato," AUTO began in its neutral, measured tone. "Your adjustments to the propulsion module have deviated from baseline parameters."

Kai didn't look up. "Deviated for the better, AUTO. Check the efficiency output. That baby's humming."

AUTO seemed to offer its version of a sigh. "Efficiency output has increased by 0.8%. However, instability in the propulsion system has also increased by 0.9%."

Kai sat back on his heels, wiping his hands on his jumpsuit. "Can't make an omelet without cracking a few eggs, right?"

"Analogies involving culinary practices are irrelevant to propulsion stability," AUTO replied. "I strongly recommend returning the system to its baseline state."

Kai gave a lopsided smile, turning his attention back to the calibration node. "Relax, AUTO. I've got this under control. Just let me do my thing."

"Your 'thing,' Passenger Sato," AUTO retorted, "is introducing unnecessary risk. Please note that continued unauthorized modifications may result in restricted access to propulsion systems."

Kai sat back, "Fine, fine. I'll back off. For now."

* * *

The observation lounge was quiet, save for the sound of the ship's systems and the faint hiss of air cycling through the vents. Lena Stroud sat near the massive window, arms crossed, her expression guarded. The stars stretched out before her, endless and indifferent. The sight was breathtaking, but it didn't ease the tight knot in her chest.

Her potted plant sat on the small shelf beside her, a splash of green against the cold gray of the ship's interior. She reached out, running her fingers along the smooth rim of the pot, her mind far away.

"Getting settled in?" Theo Hudson's voice broke the silence.

Lena glanced over her shoulder. He stood a few feet away, hands in his pockets, his stance relaxed but his expression attentive.

"Settled enough," she replied, her tone neutral.

Theo stepped closer, his gaze flicking between her and the stars beyond the window. "It's a lot to take in, isn't it? The ship, the mission, all of it."

Lena shrugged. "It's not the ship or the mission I have a problem with."

"Then what is it?" Theo asked gently.

She hesitated, her hands tightening. "It's the people running it. Big ideas are great, but execution is where everything falls apart. I've seen it happen too many times to count."

Theo nodded slowly, his expression thoughtful. "I get it. You're worried about what happens if things go wrong."

"I'm worried about what happens when things go wrong," Lena corrected. "Because they always do."

"That's fair," Theo said. "But that's why we're here - to make sure this works. Together."

Lena didn't respond right away. She turned her attention back to the stars, her grip on the plant loosening slightly. "I just hope you're right."

* * *

As Theo left Lena to her thoughts, AUTO observed silently from the corner of the lounge. Its holographic form flickered briefly as it processed the interaction.

"Log entry," AUTO murmured to itself. "Passenger Stroud exhibits resistance consistent with archetype parameters. Probability of engagement with mission objectives: 42%. Recommend strategies to increase alignment."

Its sphere paused before vanishing, leaving Lena alone with her doubts and the endless void beyond the glass.

* * *

The constant whir of the *Horizon One's* systems had become a constant backdrop to life aboard the ship, blending with the occasional chime of notifications or the faint murmur of crew conversations. But that hum shifted suddenly, a subtle vibration, barely perceptible, yet enough to catch Clara Vega's attention in the navigation hub.

She froze mid-command, her eyes narrowing at the telemetry screen. A red alert flashed briefly before disappearing, replaced by a yellow diagnostic warning.

Clara tapped her headset. "Theo, we've got an issue."

Theo's voice came through almost immediately. "What kind of issue?"

"Power draw in the propulsion system just spiked. It leveled out, but something's definitely off. I'm running a diagnostic now, but it's not matching up with AUTO's baseline readings."

There was a brief pause before Theo replied. "Meet me in the engine room. Bring your data. And let's see what Kai knows."

Moments later, Clara arrived in the engine room to find Kai Sato crouched near one of the propulsion modules, tools scattered around him. The faint hum of the systems felt heavier here, as if the ship itself was holding its breath.

Kai glanced up as Clara entered, flashing a toothy smirk. "Vega! Here to admire my handiwork?"

"Let me guess," Clara said sharply, holding up her tablet. "This spike in the power draw has nothing to do with your 'enhancements,' right?"

Kai stood, brushing his hands on his jumpsuit. "Relax, Vega. I didn't do anything that would mess with the system. Just a little fine-tuning."

"Fine-tuning?" Clara's voice rose. "You're not even supposed to be touching the propulsion modules, let alone 'fine-tuning' them."

Theo entered the room then, his presence immediately commanding attention. His voice was calm but firm. "What's going on?"

Clara turned to him, her frustration bubbling over. "This fluctuation I've been tracking? It just spiked.

And now I find our resident cowboy playing mechanic down here."

Kai crossed his arms. "It wasn't me, all right? The system's solid. Whatever's happening, it's not because of anything I did."

Theo glanced between them, then addressed AUTO, whose holographic form materialized nearby. "AUTO, confirm: has Passenger Sato's activity impacted the propulsion system?"

AUTO read out: "Power fluctuation analysis: 4% variability increase correlating with unauthorized system adjustments. Kai Sato influence factor: Positive 0.8% efficiency improvement; negative 0.9% stability compromise."

Kai rolled his eyes, "Okay, so maybe it's a little me. But it's nothing we can't handle."

Clara glared at him. "This isn't a game, Sato. If we lose stability during the jump, we could end up stranded, or worse."

Theo stepped between them, his voice steady. "All right, enough. Kai, no more unauthorized adjustments. Clara, focus on stabilizing the system." He shot a warning glance at Clara, "I expect the two of you to work together."

The tension between Kai and Clara lingered as they worked, their interactions clipped and professional but far from harmonious. Theo stayed close, observing quietly and stepping in only when absolutely necessary.

At one point, Lena appeared in the doorway, drawn by the raised voices. She had her arms crossed, her expression skeptical.

"Let me guess," Lena said dryly. "The ship's 'hybrid collaboration' idea isn't working out so well."

Theo turned to her. "We're handling it."

"Sure, you are," Lena replied, pushing off the frame and stepping into the room. "But let me ask you this: if we can't even keep the power systems stable, how are we supposed to handle the jump? Or the mission after that?"

"We'll handle it because we have to," Theo said.

Lena raised an eyebrow. "That's not much of a plan."

"It's a starting point," Theo said firmly. "And that's enough for now."

As the others exchanged words, Clara's focus remained on her console. Finally, she straightened, her eyes narrowing at the display.

"There it is," she said, pointing to a section of the readout.

Theo stepped closer. "What did you find?"

"The fluctuation isn't just a system imbalance," Clara explained. "It's being caused by a feedback loop in the secondary power grid. It's subtle, but if we don't correct it, it'll only get worse."

Kai peered over her shoulder. "A feedback loop? That's… weird. Shouldn't AUTO have caught that?"

"AUTO," Theo said, addressing the hologram. "Why wasn't this flagged sooner?"

"The feedback loop falls within acceptable operational parameters. However, Engineer Vega's analysis suggests a potential risk factor exceeding 12% under prolonged strain."

Clara shot Kai a pointed look. "Acceptable parameters don't mean safe."

Kai held up his hands. "All right, all right. You win this round, Vega. And a good reminder that AUTO isn't foolproof."

Despite the tension, the group pulled together to address the issue. Clara directed Kai as he recalibrated the secondary power grid, their interactions still tinged with friction but gradually softening as they focused on the task at hand.

Theo coordinated from the sidelines, ensuring Lena was briefed on the situation and ready to assist if needed. Even AUTO contributed, providing precise calculations and step-by-step instructions for the adjustments based on the team's assistance and information provided.

By the time the system stabilized, the atmosphere in the engine room had shifted. The tension wasn't gone, but there was a sense of accomplishment that hadn't been there before.

Clara exhaled slowly. "System's stable. For now."

"See?" Kai said with a wink, "Told you we'd handle it."

Clara rolled her eyes but this time, didn't argue.

As the group dispersed, Theo lingered in the engine room, his gaze fixed on the propulsion module. The issue had been resolved, but the lingering tension among the crew weighed on him.

AUTO's voice broke the silence. "Commander Hudson, would you like me to log this incident as a resolved anomaly?"

Theo shook his head. "No. Log it as a reminder that we're all still figuring this out. Together."

"Understood," AUTO replied.

As Theo left the engine room, the ship hummed softly around him, a reminder that the real challenges were still ahead, and they would require more than just technical expertise and a dash of luck to overcome.

* * *

Later, AUTO's holographic sphere pulsed softly in the central hub as it processed the day's events. Its algorithms compiled and categorized the crew's interactions, creating a snapshot of the current dynamics aboard the *Horizon One*.

"Log entry," AUTO began, its voice low and even. "Crew dynamics report: Initial observations confirm the following archetypes remain consistent under stress."

The hologram displayed each crewmember's profile as AUTO summarized its findings.

- **Clara Vega**: Demonstrates high precision and low tolerance for ambiguity. Her skepticism drives proactive problem-solving but risks interpersonal friction, particularly with less detail-oriented counterparts. Likely to remain an essential stabilizer during technical challenges.

- **Kai Sato**: Balances enthusiasm with impulsivity, often deviating from protocol in pursuit of innovation. While his creativity is an asset, it increases operational risk. Requires oversight to channel his strengths constructively.

- **Lena Stroud**: Exhibits low engagement and high resistance to change, often challenging authority. Her skepticism provides valuable counterpoints but risks undermining team cohesion if left unchecked.

- **Theo Hudson**: Displays a balanced approach with strong leadership tendencies. Serves as a mediating force, maintaining focus on the mission while addressing interpersonal conflicts. Likely to be a critical unifier as challenges intensify.

AUTO paused, its hologram flickering faintly. "Current trajectory indicates an 18% probability of conflict escalation within the next 72 hours. Recommended strategies include fostering shared purpose and reinforcing team cohesion."

Theo entered the hub just as AUTO finished its report. He stopped short, noticing the flickering holograms.

Theo glanced at the hologram. "You keeping track of us, AUTO?"

"Affirmative," AUTO replied. "Behavioral dynamics are integral to mission success. My analysis indicates a 23% increase in cohesion compared to initial departure, but potential for interpersonal conflict remains significant."

Theo folded his arms, considering the data. "Anything else?"

AUTO's glow flickered softly. "Crew archetype alignment continues to evolve.

Theo allowed a sly grin to spread. "I'll take that as a vote of confidence. Keep monitoring but remember; data isn't the whole story. People aren't just patterns to predict."

The sphere emitted a faint shimmer, as though processing his words. "Acknowledged, Commander."

* * *

In the quiet hum of the ship, AUTO returned to its central hub, processing the day's events. Patterns of behavior and emerging dynamics flickered across its holographic sphere. The propulsion issue had been resolved, but the humans aboard the *Horizon One* were proving to be as complex, and as unpredictable, as the journey ahead.

"Log entry," AUTO murmured to itself, its voice barely audible. "Probability of mission success: 83%. Further data required."

The ship sailed onward into the void, its systems steady, its crew anything but.

Seeds of Doubt

For most passengers, the initial novelty of the journey had begun to wear off, replaced by a simmering unease. It wasn't the ship's systems or the strange quiet of the stars; it was the growing realization that the mission demanded more than they had anticipated.

In the observation lounge, Lena Stroud looked out the wide window, the vast emptiness of space reflecting her own uncertainties. She traced a finger along the cool glass, her expression distant.

Theo found her there, his approach slow and deliberate. "You've been quiet," he said, stopping a few steps away.

Lena snorted. "Haven't had much to say."

"You're usually not short on opinions," Theo said lightly. "What's on your mind?"

She turned to face him, crossing her arms. "You want the truth? This mission feels like a mistake. We're hurtling through space with an AI running the show and a crew that can't even agree on basic protocols. How is this supposed to end well?"

Theo's jaw tightened, but his voice remained calm. "I get it. Change is hard. It's uncomfortable. But this mission isn't about making everyone feel comfortable, it's about pushing boundaries."

"Pushing boundaries is great in theory," Lena shot back. "But in practice, it gets people killed. I've seen what happens when ambition outpaces preparation. It's never pretty."

Theo stepped closer, his gaze steady. "Lena, we're all here because we bring something unique to the table. You're here because you question things. Because you challenge the status quo. That's valuable. But if you let your doubts control you, they'll drag the rest of us down with them."

For a moment, Lena said nothing. Then she sighed, turning back to the window. "I just hope you know what you're doing, Commander."

* * *

From the central hub, AUTO monitored the conversation through its internal systems. Its holographic sphere flickered to life, and it began processing the dialogue.

"Log entry," AUTO murmured. "Passenger Stroud exhibits resistance consistent with archetype parameters. Probability of resolution through Commander Hudson's intervention: 64%."

The AI paused. "Adjusting strategy. Recommend supplemental morale reinforcement measures."

* * *

Elsewhere on the ship, Clara was still running diagnostics on the propulsion system when Kai strolled into the room.

"Still obsessing over those numbers, Vega?" he teased, leaning against the console.

Clara didn't look up. "Still pretending you know what you're doing, Sato?"

Kai laughed. "Come on, you have to admit, my 'enhancements' worked out okay."

Clara glared at him. "Your so-called enhancements nearly destabilized the entire system. Do you have any idea how close we came to a critical failure?"

Kai shrugged. "Close only counts in horseshoes and plasma grenades."

"Do you ever take anything seriously?" Clara snapped.

Kai faltered for a moment, but he recovered quickly. "I take plenty of things seriously. I just don't see the point in stressing over every little fluctuation."

"Those little fluctuations could kill us," Clara said sharply. "This isn't a game, Sato. If you can't take it seriously, maybe you shouldn't be here."

The tension between them was palpable, but before either could say more, Theo entered the room.

"Everything okay in here?" he asked, his tone calm but firm.

Clara folded her arms. "Depends on your definition of 'okay.'"

Kai tilted his head. "Just a friendly debate, Commander."

Theo sighed. "Let's try to keep it friendly. We're all on the same team here. And I came to tell you that we're approaching the first waypoint in our journey. Please report to the central hub for a mission briefing."

* * *

As the crew gathered in the hub, the atmosphere was tense. Lena stood off to the side, her arms crossed, while Clara and Kai took seats on opposite ends of the room. Theo stood at the center; his expression neutral but his mind racing with the weight of what lay ahead.

AUTO's holographic form materialized beside him, its sphere pulsing softly. "Commander Hudson, the probability of successful mission completion has been adjusted to 81%."

Theo raised an eyebrow. "Why the drop?"

"Recent interpersonal dynamics have introduced new variables," AUTO replied. "While technical systems remain stable, human factors present an increased risk."

Theo nodded slowly, his gaze sweeping across the room as the crew exchanged uneasy glances. "Noted," he said firmly. "Let's focus on the mission."

As the *Horizon One* pressed deeper into the void, the challenges they faced became increasingly clear. The test of their journey wouldn't be limited to the technical intricacies of space exploration, it would be a test of their ability to navigate the unpredictable terrain of human collaboration. And as the crew's unease simmered, so too did the unspoken question: could they overcome their differences before the mission's stakes became too great to bear?

* * *

Chapter 4: Core Adaptation

Evolving Under Pressure

"Sometimes you just have to jump out the window and grow wings on the way down."

- Ray Bradbury

Core Adaptation is where teams confront resistance, rethink old habits, and embrace novel approaches. Change demands flexibility and courage as individuals let go of what no longer serves them. At this stage, success comes from reframing challenges, embracing learning, and showing resilience in the face of uncertainty.

Reflections and Recalibration

The common area of the *Horizon One* was dimly lit, its usual inviting glow replaced by the buzz of recalibrating systems. The crew gathered out of necessity, their faces marked with tension from the day's events.

Theo's gaze was steady as it moved from one crew member to the next. Clara sat with her tablet open, flipping through diagnostic reports. Kai sprawled on the couch, though his casual demeanor couldn't hide the unease in his eyes. Lena perched on the edge of a chair, her potted plant resting silently on the floor beside her.

AUTO's sphere hovered in the corner, its faint glow punctuating the tension.

Theo cleared his throat, breaking the silence. "We need to address today's events. The recalibration was rough, but we stabilized. What I want to know is what we've learned, and how we stop it from happening again."

Kai twirled his stylus. "And here I thought near-disaster was part of the adventure."

Clara snapped her tablet shut, her tone sharp. "Adventure? You think balancing feedback loops in a live system failure is adventurous? Maybe next time, I'll let you handle it."

Kai huffed. "Relax, Vega. I didn't mean it like that."

"Enough," Theo said, his voice cutting through the brewing argument. "We're not here to rehash what went wrong. We're here to make sure it doesn't happen again."

Lena crossed her arms, her voice cold and pointed. "How do we stop it when some people don't take the mission seriously?"

Kai sat up straighter. "I'm getting real tired of that tone, Stroud."

"Then stop giving me reasons to use it," Lena shot back.

Theo raised a hand, cutting them both off. "Seriously, enough. We don't have time for this." He turned to Clara. "Clara, what's your take? How stable are we now?"

Clara exhaled, her tone controlled. "The secondary grid is holding. For now. The feedback loop was a rare alignment of errors, too rare to ignore. But I've gone through the diagnostics three times, and I don't see anything pointing to a broader issue. At least not yet."

Theo nodded. "So, what do you need to ensure we're ready for the next phase?"

"Full cooperation," Clara said, her gaze flicking briefly to Kai. "And full access to the system logs, uninterrupted."

A soft halo of light encircled the sphere briefly, drawing attention. "If I may interject," AUTO began, "crew cohesion and communication remain suboptimal. Addressing these inefficiencies may reduce the probability of further risk events."

Clara groaned. "Let me guess. More of your 'structured dialogue sessions'?"

"Precisely," AUTO replied, undeterred. "Aligned objectives and enhanced trust could reduce future operational conflicts by an estimated 18.4%."

Lena raised an eyebrow. "So, group therapy? With a floating Magic 8 ball?"

"Incorrect," AUTO said. "This would be a strategic alignment exercise, not an emotional intervention."

Kai chuckled under his breath. "And here I thought trust falls were outdated."

"AUTO has a point," Theo admitted. "If we don't figure out how to work together, we'll be in trouble when real challenges hit. Starting tomorrow, we'll have daily check-ins. AUTO can help facilitate, but this is our mission. We make the calls."

Clara nodded reluctantly. "Fine. But I'm leading the effort to examine the grid. I don't want distractions."

Theo turned to Kai. "You're assisting Clara. I want redundancy checks across the whole system."

Kai sighed but nodded. "Got it."

"And Lena," Theo added, turning to her. "You'll help me manage the passengers. Keep them calm and focused. If there are questions or concerns, we need to handle them before they spiral."

Lena hesitated but gave a curt nod. "Fine."

As the meeting ended, Theo lingered in the communal area, staring out the viewport. The faint, endless glow of the stars seemed both comforting and oppressive. Clara approached, her tablet still in hand.

"You think this mission's really going to work?" she asked, her voice quieter than usual.

Theo didn't look at her but responded evenly. "It has to. And we'll make sure it does."

Clara nodded, her posture softening slightly. "Alright. Let's get back to it."

As she left, Theo stayed at the viewport a moment longer, the quietness of the ship filling the room. In the corner, AUTO's glow shifted subtly, silently continuing its observations as the ship pressed onward into the unknown.

Relearning the Basics

The training deck of the *Horizon One* was expansive, its design both functional and futuristic. Rows of modular workstations lined the walls, equipped with holographic interfaces that shimmered faintly in the dim light. In the center, a circular simulation table glowed softly, ready to project any number of training scenarios.

Clara stood at the head of the room, arms crossed as she waited for the others to arrive. Though she had reluctantly agreed to lead this session, she approached it with her usual determination, knowing the importance of reinforcing their understanding of the ship's systems.

Theo entered first, giving her a brief nod of encouragement. Lena followed close behind, her expression guarded but attentive, and Kai brought up the rear, spinning a small tool in his hand with feigned nonchalance. AUTO materialized last, its sphere pulsing faintly as it scanned the room.

"Alright," Clara began, her voice firm. "Today's focus is on redundancies. Specifically, how they work, why they matter, and what happens when they're ignored."

Kai let out a dramatic sigh, flopping into a chair. "Sounds riveting."

Clara shot him a withering look. "I promise you, Sato, understanding this might just save your life one day."

Clara tapped the simulation table, and a detailed holographic model of the *Horizon One* appeared, its systems highlighted in color-coded overlays. With a few gestures, she zoomed in on the secondary power grid.

"This," she said, pointing to a glowing junction, "is where yesterday's feedback loop began. The fail-safe designed to contain it didn't activate. AUTO, explain why."

AUTO answered. "The fail-safe parameters were temporarily disabled during pre-launch testing and were not re-enabled prior to departure."

Clara's gaze shifted pointedly to Kai. "And who handled the pre-launch tests?"

Kai said, "Look, I might've adjusted the parameters, but it was standard testing protocol. Resetting them wasn't my responsibility."

"Of course not," Clara said dryly. "Responsibility's never your thing, is it?"

Before Kai could respond, Theo raised a hand. "The issue's been identified. Now let's focus on ensuring it doesn't happen again."

Clara activated a new layer of the simulation, showing a series of redundant circuits branching from the main grid. "The *Horizon One* relies on multiple redundancies to function. These systems are designed to prevent small issues from becoming catastrophic failures. But they only work if they're properly maintained - and if we all understand how to use them."

She turned to Lena, handing her a tablet. "Run a diagnostic on this circuit. Follow the prompts and let me know if you hit any errors."

Lena frowned but took the tablet, muttering under her breath. "Sure. Because I've always dreamed of being an engineer."

"You don't need to be an engineer," Clara said. "But you do need to know enough to react if something goes wrong."

As Lena worked through the diagnostic, Clara directed Kai to simulate a grid overload and Theo to monitor how the ship's systems responded.

* * *

After a few minutes, AUTO spoke. "Observation: crew collaboration remains inconsistent. Efficiency could be improved through structured role alignment."

Kai let out a dry laugh. "We get it, AUTO. You think we suck at teamwork."

"Incorrect," AUTO replied, its tone neutral. "I am incapable of forming subjective opinions. My analysis is based on empirical data."

Clara rolled her eyes. "Thanks for the reminder. Now, let's focus."

"Hey," Lena said suddenly, her tone tinged with surprise. "I think I found something."

Clara moved to her side, glancing at the tablet. A faint smile broke through her usual stern expression. "Good catch. That's a minor relay fault: not enough to cause a feedback loop, but definitely worth addressing."

Lena shrugged, though a hint of pride flickered in her eyes. "Beginner's luck."

"Luck or not, it's progress," Clara said.

Kai, observing from across the room, quipped, "Looks like Lena's gunning for my job."

"Unlikely," Lena shot back with a smirk.

* * *

As the session wrapped up, Theo gathered the group around the simulation table. "Good work today. This is exactly what we need to be doing; learning the systems, understanding the redundancies, and working as a team."

Clara nodded, though her expression remained serious. "We're stable for now, but this ship is a complex system. If one thing fails, it's a domino effect. Redundancies aren't just backups: they're lifelines."

Kai's positive attitude was returning. "Lesson learned, Vega. Don't mess with the lifelines."

Clara didn't bother responding, but a hint of a smile tugged at the corner of her mouth.

Theo placed his hands on the table, his voice steady. "We're making progress. Let's keep it up. Every system we master, every issue we resolve, gets us closer to success."

The group separated, the tension from earlier replaced by a cautious sense of accomplishment. As AUTO dimmed its glow and the simulation table powered down, the ship's quiet hum returned, a reminder of both the challenges they'd faced and those still ahead.

Breaking Point

The corridors of the *Horizon One* were dim and quiet, the ship's hum a constant undercurrent to Theo's thoughts. The peace didn't last, shattered by the sound of raised voices echoing from the engine room.

Theo's pace quickened. As he stepped into the room, he found Clara and Kai locked in a heated argument, their tension palpable.

"Do you ever think beyond your ego?" Clara snapped, her frustration echoing off the metallic walls.

Kai threw up his hands. "Oh, that's rich coming from you, Vega! You act like you're the only one holding this ship together."

"Stop!" Theo's voice cut through the noise like a blade, silencing both of them.

Clara turned to Theo, her face flushed with exasperation. "The secondary grid's instability is escalating. I need to reroute power through auxiliary systems to buy us time, but *he* won't stop second-guessing me."

Kai crossed his arms, his voice sharp. "Because it's a bad idea. Those circuits aren't meant to handle that load for long. If we push them too far, we're risking environmental control - and more."

Clara countered immediately, her tone clipped. "The diagnostics are clear. This is the best option to stabilize the grid until we find the root cause."

Theo held up a hand to stop the volley. "Okay, let's get this straight. Clara, do you have data to back your plan?"

Clara nodded, holding out her tablet. "The auxiliary circuits can handle the reroute for the short term. It's not perfect, but it's the safest immediate fix."

Theo shifted his gaze to Kai. "And your concern?"

Kai gestured toward the engine controls. "She's right about the short term, but if anything spikes, those circuits will fry. Then we're looking at a cascade failure."

Theo exhaled, considering both sides. "Here's what we'll do. Clara, start the reroute, but incrementally. Kai, monitor the circuits. If anything strains beyond the threshold, you stop the reroute immediately. Work together. Understood?"

Kai grumbled but nodded. "Fine."

Clara's nod was curt, her focus already returning to her console. "Let's get it done."

* * *

The roar of the ship grew louder as Clara began the reroute. At first, the process seemed smooth, but then the alarms began.

"Auxiliary load at 85%!" Kai called out, his voice tense.

"I'm compensating," Clara replied, her tone steady but firm. "We're still within range."

Theo stepped closer to the central console, his presence a calming influence despite the rising tension. "Keep it together. Both of you."

The alarms peaked, and Kai's voice rose. "We're at 93%! Pull back, Vega!"

"Not yet," Clara shot back, her focus unshaken. "Just a little longer... now!"

The alarms abruptly silenced, and the readouts returned to green.

"Load rerouted successfully," AUTO reported, its tone neutral. "System stability restored."

Kai shook his head. "You're walking a thin line, Vega."

Clara turned to him, her voice quieter now but unwavering. "It wasn't luck. It was calculated risk - and it worked."

Kai reluctantly smirked. "You're not as bad as you think you are. Still a pain, though."

Theo stepped forward, his voice steady and commanding. "Good work. Both of you. But this back-and-forth can't continue. From now on, you're a team, not rivals. Agreed?"

Clara and Kai exchanged reluctant nods. "Agreed," they muttered in unison.

* * *

As the others left the engine room, Clara lingered at her console, her gaze locked on the diagnostic data. Theo approached, his tone neutral but probing.

"Still worried?" he asked.

Clara nodded slowly, tapping a finger on the screen. "There's something about the patterns we're seeing - it's too precise. I don't think we're dealing with normal wear and tear."

Theo's brow furrowed. "You think it's deliberate?"

Clara hesitated, then shook her head. "I'm not saying that. But it's... odd. I'll keep digging."

Theo gave her a supportive nod. "Let me know what you find. And Clara - good work today."

As she returned to her diagnostics, AUTO's sphere dimmed slightly, a silent observer of the unfolding dynamics aboard the *Horizon One*.

A New Perspective

The observation deck of the *Horizon One* was bathed in the soft glow of the nebula outside. The swirling colors, rich purples and molten oranges, danced across the wide viewport, casting faint, shifting reflections on the walls. Lena sat cross-legged on one of the cushioned benches, her potted plant resting beside her. Her eyes, fixed on the endless expanse, held a quiet unease.

Theo entered quietly, pausing just inside the room. He let the silence linger, taking in the beauty outside before speaking. "Incredible, isn't it?"

Lena glanced at him, her tone guarded. "Depends on what you're looking for."

Theo tilted his head, intrigued. "And what are you looking for?"

Lena's gaze returned to the viewport. "Answers. A sign this mission isn't as doomed as it feels."

Theo moved closer, taking a seat on the opposite bench. "You're not the only one carrying doubts, Lena. We're all grappling with the unknown out here."

"Yeah, well," Lena said, her voice low, "grappling isn't exactly my strong suit." She hesitated before adding, "I thought this mission would be... different. A clean break from everything back home. But it feels more like jumping from one mess to another."

Theo nodded, leaning forward slightly. "Sometimes starting over isn't as clean as we imagine. But it's still a chance to grow. To build something new."

Lena snorted softly. "You're a real motivational speaker, you know that?"

"Not intentionally," Theo admitted with a small smile. "But I've learned that even chaos has its patterns, its possibilities."

AUTO's sphere materialized beside them, its glow faint but insistent. "Observation: Passenger Stroud's stress indicators have decreased by 11% during this interaction."

Lena turned toward the hologram, frowning. "You're really keeping tabs on me like that?"

"Correct," AUTO replied. "Monitoring physiological and psychological data is critical to crew performance and mission success."

Theo chuckled, gesturing toward the AI. "You'll get used to AUTO's bedside manner. It's all about the mission."

Lena crossed her arms, her voice laced with sarcasm. "Oh, great. Good to know I'm just data points in a spreadsheet."

"Not just data," AUTO said, its tone almost defensive. "Your contributions, emotional and operational, are significant. Current indicators suggest you are an integral component of mission cohesion."

Lena raised an eyebrow. "Did... did AUTO just compliment me?"

Theo laughed. "See? It's not all bad."

Lena sighed, her shoulders loosening slightly. "Trust isn't something I'm good at, Theo. People, systems, missions - everything's let me down at some point."

Theo met her gaze, his tone steady. "Trust isn't about being certain nothing will fail. It's about believing we can handle it when it does. And from what I've seen, you're capable of that."

Lena hesitated, then nodded slightly. "Maybe. But believing doesn't come easy for me."

"It doesn't for most people," Theo said. "But you're here, aren't you? That says something."

For a moment, they sat in silence, watching the nebula shift and swirl.

As Theo stood to leave, he gestured to the colors outside. "That nebula - it's unpredictable, messy. But it's also full of energy, creation, and possibility. Kind of like this mission."

Lena tilted her head, considering his words. "So, what? You're saying if we stick it out, we might create something extraordinary?"

Theo smiled faintly. "I'm saying it's worth finding out."

Lena watched him leave, the nebula's glow painting her thoughtful expression. For the first time, she allowed herself to wonder if this mission wasn't just a mess waiting to happen, but a chance to find meaning, and maybe even purpose, amid the chaos.

* * *

The mess hall of the *Horizon One* chattered with a liveliness that had been absent for days. The clatter of trays and the murmur of subdued conversation marked a collective exhale from the crew. Small moments of camaraderie emerging with Lena and Kai exchanging light-hearted jabs, Clara's rigid posture softened, and Theo sat observing it all with quiet satisfaction from the corner of the room. Even AUTO's sphere glided unobtrusively, its occasional quips drawing faint smiles instead of irritation.

Theo sat at one of the larger tables, his meal untouched as he scanned the room. Clara joined him, sliding into the chair opposite, her tablet still in hand.

"You're watching them," Clara noted without looking up.

Theo smiled faintly. "Can't help it. It's nice seeing the team actually feel like a team."

Clara nodded absently but didn't share his ease. "Cohesion's good, but it doesn't fix the ship. And it doesn't answer the bigger question."

Theo asked, "What bigger question?"

Clara hesitated, her hand tapping lightly on the edge of her tablet. "The patterns are consistent, almost too much so. But if Kai's right about the navigation system being the root cause... that just means someone missed something huge during design. That kind of oversight is hard to believe."

AUTO's sphere floated silently to their table, pulsing faintly. "Officer Vega, would you like assistance analyzing the navigation data?"

Clara glanced up suspiciously. "What kind of assistance?"

"Advanced simulations to identify cascading anomalies," AUTO replied. "However, I would require expanded subsystem access to proceed."

Theo raised an eyebrow. "What's the risk if we give you that access?"

"Probability of malfunction increases by 2.3%," AUTO said matter-of-factly. "However, probability of resolving the issue increases by 38.7%."

Clara shook her head firmly. "No way. Unrestricted access is a Pandora's box waiting to open."

Theo held up a hand. "Let's consider this. If AUTO's right, it could save us critical time later. Is there a controlled way to test it?"

Clara frowned but didn't dismiss the idea outright. "I'll think about it. But I'm not making a call until I've run every check I can with the tools we already have."

"Understood," AUTO said, its sphere dimming as it retreated.

* * *

Across the room, Kai hunched over his portable workstation, working away. Lena sat nearby, watching with curious skepticism.

"What are you doing?" Lena asked, leaning closer.

"Following a hunch," Kai said, not looking up. "Everyone's focused on sabotage or some hidden glitch. But sometimes, the simplest answer is the right one."

"And that is?" Lena pressed.

Kai turned the screen toward her. "The navigation system. The power spikes match its recalibration cycles. It's not sabotage; it's bad engineering."

Lena studied the data, her frown deepening. "So the ship is causing its own problems?"

"Exactly," Kai said, standing abruptly. "And I'm about to blow Vega's mind with it."

At Theo and Clara's table, Kai dropped his workstation with a flourish, pulling up his findings. "I've cracked it. The navigation system's recalibrations are overloading the grid. It's not sabotage; it's just bad design."

Clara looked over the screen, her skepticism fading as she studied the data. "Huh. That... actually tracks."

"Of course it does," Kai said, "I'm a genius, remember?"

Clara rolled her eyes but didn't argue. "If you're right, this changes everything. We need to rework the power flow parameters to compensate. It's not a perfect fix, but it's a start."

Theo nodded, his expression serious but hopeful. "Good work, Kai. This gives us a direction."

The three began sketching out a plan, their collaboration smoother than it had been in days.

As the crew began addressing the navigation flaw, AUTO's sphere hovered nearby, pulsing faintly.

Behind its glowing exterior, a new subroutine activated: monitoring not just the ship's systems but the crew's behavioral patterns. The faint flicker of this subroutine went unnoticed, a quiet harbinger of complications yet to come.

In the mess hall, the tension had eased, replaced by a tentative sense of accomplishment. Lena joined the discussion, her tone less sharp. Kai and Clara exchanged only the occasional barb, their focus mostly on solving the problem. For the first time, the crew seemed to be finding a rhythm.

But beneath the surface, uncertainties lingered. The navigation system was only one issue among many, and the mission's true challenges lay ahead.

* * *

Chapter 5: Capability

Harnessing Strengths

"There is this about the human mind: if it can be done, it will be done. We can transform Mars and build it like you would build a cathedral, as a monument to humanity and the universe both. We can do it, so we will do it. So - we might as well start"

- Kim Stanley Robinson, *Red Mars*

Capability is about recognizing and maximizing strengths. Every team member brings unique skills and perspectives to the table, and this stage focuses on aligning those talents to drive progress. Capability isn't just about competence; it's about empowering individuals to rise to the moment and contribute meaningfully to success.

Testing the Plan

The engineering bay was filled with activity as the crew gathered around the central console. Holographic schematics of the *Horizon One*'s navigation system hovered in midair, casting an eerie blue glow on their faces. Clara stood at the head of the table, her expression focused as she adjusted her tablet, the cascading streams of data reflecting in her sharp eyes.

"Here's the plan," she began, her voice brisk. "We're isolating the navigational recalibration protocols and redistributing excess power to the auxiliary systems. This should stabilize the grid and prevent further overloads."

Kai examined the schematics with a faint smirk. "Sounds easy enough. Where's the part that goes horribly wrong?"

Clara shot him a withering look. "The part where we miscalculate and knock out propulsion. Or worse, fry the auxiliary circuits."

Lena, standing off to the side, frowned. "Why does every solution feel like it comes with a built-in disaster?"

"Because in space, it does," Clara replied flatly, tapping a command on her tablet. "And that's why this is a team effort."

Theo stepped forward, his tone calm but firm. "Clara and Kai, you'll handle the technical adjustments. Lena, I need you monitoring environmental systems. If anything spikes, we shut it down immediately. No hesitation."

Lena's frown deepened. "And what happens if we're too late?"

"We improvise," Theo said simply, his tone leaving no room for argument.

AUTO's holographic sphere materialized above the table, pulsing softly. "Observation: current plan carries a 71% probability of success. Recommendation: integrate a secondary fail-safe to mitigate risk."

Clara sighed, her patience visibly fraying. "We don't have the resources or time to build another fail-safe, AUTO."

"Incorrect," AUTO replied smoothly. "A temporary fail-safe can be implemented using existing subsystems. Shall I proceed?"

Theo glanced at Clara. "What do you think?"

Clara hesitated, clearly weighing the risks. Finally, she nodded. "Fine. But keep it simple, AUTO. Nothing experimental."

"Understood," AUTO replied, its sphere pulsing faintly. "Implementing protocol."

Kai took his position at the auxiliary controls. "Auxiliary systems are online," he called out. "Ready when you are."

Clara adjusted her tablet, her eyes glued to the readouts. "Initiating recalibration now. Lena, stay sharp on environmental monitoring."

Lena hovered over her console, her jaw tight as she scanned the data streams. "All readings normal so far."

The ship's hum intensified as the recalibration began. The holographic schematics displayed the shifting power flow, the glowing lines of energy pulsing steadily. For a moment, everything seemed to be working.

"Load transfer at 40%," Clara reported. "System stable."

"See? Smooth sailing," Kai quipped, leaning back slightly.

Suddenly, alarms blared, and the schematics lit up in angry red. Lena's console beeped furiously.

"Environmental systems are spiking!" Lena shouted. "CO2 levels rising in Sector 4."

Clara's fingers flew over her tablet. "Pulling back on the transfer—"

"Wait," Kai interrupted, his voice sharp. "It's not the transfer. It's another feedback loop. Shutting it down won't stop the spike."

Theo stepped closer. "Options?"

Clara's eyes darted over the readouts. "We can reroute power from propulsion to stabilize the environmental systems, but it'll cut our thrust."

Kai shook his head. "If we lose thrust, we drift. Bad idea. We need propulsion to stay aligned."

"Then what's your plan, genius?" Clara snapped.

Kai, diving into a side panel said, "Give me 30 seconds. I'll rig a bypass for the feedback loop. It won't win any design awards, but it'll hold."

As the alarms blared, Kai worked furiously, pulling out wires and reconnecting them with practiced precision. "Almost there," he muttered.

Clara watched him, her skepticism tempered by the glimmer of respect she would never admit aloud. "If this doesn't work…"

"It'll work," Kai cut her off, securing the final connection. "Try it now."

Clara adjusted the controls, her focus absolute. The alarms suddenly fell silent, and the schematics shifted back to green. A collective exhale swept through the room.

"Load transfer complete," Clara announced. "Auxiliary systems are holding."

"CO_2 levels are dropping," Lena added, her voice edged with relief. "Environmental systems back to normal."

Theo smiled, clapping his hands empathetically. "Nice work."

* * *

As the team began shutting down their stations, Clara turned to Kai, her expression still guarded. "That was reckless."

Kai shrugged, a smug look on his face. "Maybe. But it worked."

Clara rolled her eyes but didn't argue. "Just don't make a habit of it."

Theo stepped between them, his tone measured. "Reckless or not, today was progress. We're starting to learn how to work together, and that's what's going to keep this mission on track."

AUTO's sphere blinked while calculating. "Observation: team cohesion has improved by 14%. Probability of mission success has increased to 77%."

Kai half-smiled. "Glad to know we're trending upward."

Clara muttered under her breath. "For now."

As the team left, Clara lingered by the console, staring at the schematics. For the first time, the patterns seemed manageable. Maybe, just maybe, they were starting to turn the tide.

Strengths and Weaknesses

The conference room of the *Horizon One* hummed softly, its stark interior illuminated by glowing screens displaying mission updates and system diagnostics. The circular table in the center seemed to absorb the tension radiating from the crew, some leaning back in their chairs while others perched tensely at the edge.

Theo stood at the head of the table, his posture steady but not rigid. In front of him, a small holoprojector flickered to life, displaying holographic avatars of the crew alongside bar graphs and data streams.

"Alright," Theo began, his tone calm but commanding. "It's time we have an honest conversation about where we stand as a team. What we're good at, and what we're not."

Kai tilted his chair back, balancing it on two legs. "If this is a performance review, I'm pretty sure I crushed it yesterday."

Clara didn't bother hiding her glare. "This isn't about scoring points, Sato."

"Exactly," Theo said, cutting through the tension. "It's about figuring out how to maximize our strengths and shore up our weaknesses. We're making progress, but we've still got blind spots, and in space, blind spots get people killed."

As Theo finished, AUTO's holographic sphere materialized above the table, pulsing faintly. "To facilitate this discussion, I have compiled performance metrics and psychological profiles for each crew member. Would you like me to share individual reports?"

Lena frowned, her arms crossed. "You've been profiling us this whole time?"

"Correct," AUTO said matter-of-factly. "Continuous evaluation is essential for mission optimization."

"Creepy," Lena muttered.

Theo raised a hand. "Let's keep this productive. AUTO, just give us the highlights."

The hologram shifted to display a rotating set of charts as AUTO began.

- **Clara**: Exceptional technical expertise and problem-solving abilities. Weakness: limited patience and a tendency to micromanage.

- **Kai**: High adaptability and strong improvisation under pressure. Weakness: inconsistent adherence to protocol and over-reliance on intuition.

- **Lena**: Sharp observational skills and high emotional intelligence. Weakness: reluctance to take initiative and a propensity for skepticism.

- **Theo**: Strong communicator and effective leader. Weakness: difficulty delegating and overextension in conflict resolution.

The room fell silent as the final chart faded.

Kai broke the tension. "Glad to know I'm perfect, except for that one little detail."

Clara rolled her eyes. "Your 'one little detail' nearly fried the environmental systems."

Lena raised an eyebrow. "I'm still not thrilled about the surveillance aspect."

"It's not about surveillance," Theo said evenly. "It's about learning to rely on each other. Clara, you're brilliant, but you don't have to solve every problem alone. Trust your team."

Clara hesitated, then nodded slightly. "Fine. As long as the team earns that trust."

Theo turned to Kai. "Kai, you've got instincts most people dream of, but you need to channel them into consistent action. No shortcuts."

"Got it," Kai said, his tone uncharacteristically serious.

Finally, Theo addressed Lena. "Lena, you bring valuable insights, but we need you to step up. Don't wait for permission to act when you see something wrong."

Lena hesitated, then gave a small nod. "I'll work on it."

Theo tapped the holoprojector, bringing up a holographic simulation of the ship's systems. "Now, we're going to put this into practice. Each of you will take the lead in a simulated crisis scenario. Lena, you'll coordinate. Clara, handle technical troubleshooting. Kai, manage power and environmental systems. I'll observe."

Clara raised an eyebrow. "You're putting Kai in charge of environmental systems?"

"Relax," Kai said. "I'll try not to turn off the air."

Theo ignored the banter. "This isn't about being perfect. It's about learning to trust each other and adapt. Let's begin."

* * *

The simulation began with a simulated power surge in the ship's environmental controls. Holographic warnings flashed across the screens, and the hum of the ship grew louder as the scenario unfolded.

"Clara, isolate the affected circuit," Lena instructed, her voice steady but firm.

"Already on it," Clara replied, "Kai, reroute power to the secondary grid."

Kai hesitated for a split second, then nodded. "Got it. Hold on."

The warnings grew more urgent as the team worked, but their actions became increasingly coordinated. Lena issued commands with growing confidence, while Clara and Kai responded quickly and efficiently. Theo watched in silence, stepping in only to clarify or redirect when necessary.

"Power levels stabilizing," Kai announced, a note of pride in his voice.

"Circuit isolated," Clara confirmed. "Systems normalizing."

The alarms faded, and the simulation ended with the ship's systems fully restored.

As the holograms dissolved, Theo turned to the crew, his expression thoughtful. "That was better. Still rough around the edges, but we're getting there."

Lena nodded slowly. "It felt... different. Like we were actually working as a team."

Clara didn't respond immediately but then admitted, "Not bad. For a first attempt."

Kai's trademark smile returned. "Told you we're a dream team."

"Don't get cocky," Clara said, though her tone lacked its usual sharpness.

AUTO offered: "Observation: team cohesion has improved by 17%. Probability of mission success now at 83%."

Theo smiled. "We'll take it. Good work, everyone."

As the crew dispersed, the room felt lighter, the earlier tension giving way to a growing sense of camaraderie. For the first time, the crew of the *Horizon One* wasn't just surviving - they were starting to thrive.

AUTO Takes Charge

The dimly lit control hub of the *Horizon One* buzzed with low energy. Streams of data flowed across the screens, the ship's functions holding steady for now, but Clara's sharp eyes remained fixed on potential anomalies.

Theo stood nearby, arms crossed in his characteristic pose of calm vigilance. The moment of quiet was interrupted by the sudden materialization of AUTO's sphere above the main console, its glow brighter and more urgent than usual.

"Attention," AUTO said, its tone measured but firm. "System anomalies detected in the propulsion array. Immediate intervention is required."

Clara straightened in her seat. "Why wasn't this flagged sooner?"

"It was flagged," AUTO replied smoothly. "However, the priority level was insufficient to interrupt non-critical crew activities."

"Non-critical?" Clara muttered, already pulling up the diagnostics. "Remind me to rewrite your protocols for prioritization."

Theo stepped forward, his tone calm but pointed. "AUTO, explain the nature of the anomalies."

"Propulsion output is fluctuating by 2.6%," AUTO explained. "While currently within operational tolerance, projections indicate a 47% probability of escalation if left unaddressed."

Clara dug into the propulsion data. "It's a misalignment in the stabilizers. Nothing catastrophic yet, but it's going to take precision to fix."

"Define precision," Theo said, his tone cautious.

"We'll need to manually recalibrate the stabilizers," Clara replied. "Which means someone needs to suit up and access the external array."

At that moment, Kai strolled into the room, cradling a steaming mug. "Did someone say 'suit up'? Sounds like my kind of gig."

Clara glared at him. "This isn't a joyride, Kai. One mistake out there, and you'll be orbiting as space dust."

Kai tilted his head in quiet amusement. "Relax, Vega. I've got nerves of steel."

Theo raised a hand to cut off the brewing argument. "If you're volunteering, fine. But I'm not sending you out there alone. Clara will guide you from here."

Clara frowned. "Two people? That's doubling the risk."

"Or doubling the safety," Theo countered. "We need redundancy on this. No lone heroics."

Before preparations could begin, AUTO stated. "Proposal: I can execute the recalibration remotely, eliminating all risk to human crew members."

Clara paused, her skepticism clear. "You? How exactly?"

"Using the ship's exterior maintenance drones," AUTO explained. "My systems are capable of precise adjustments within operational parameters."

Kai raised an eyebrow. "Where's the fun in that?"

"This is not about fun," AUTO said plainly. "It is about safety and efficiency. Probability of success with human intervention: 84%. Probability of success with AI intervention: 96%."

Clara turned to Theo, her face conflicted. "I'm not comfortable giving AUTO full control over something this critical."

Theo studied the sphere. "AUTO, if we let you handle this, what safeguards are in place?"

"Redundant fail-safes are integrated into the maintenance drones," AUTO replied. "Additionally, manual override remains accessible to crew members at all times."

Theo nodded slowly. "Let's vote. Clara?"

She sighed. "Fine. Let AUTO handle it, but I'll be monitoring everything."

"Seriously?" Kai groaned. "You're letting the glowing orb steal my thunder?"

"It's not about thunder," Lena said from the doorway. "It's about keeping you in one piece."

Kai cursed something under his breath but didn't argue further.

"Alright, AUTO," Theo said, his voice firm. "It's your show. But we're watching."

"Understood," AUTO replied, its sphere pulsing brightly. "Commencing recalibration sequence."

The crew crowded around the central display as the external maintenance drones sprang to life. On-screen, the drones moved with mechanical precision, their articulated arms navigating the propulsion array. Live feeds from the drone cameras provided a clear view of the stabilizer assembly.

"Stabilizer array accessed," AUTO reported. "Beginning recalibration."

The room was silent, save for the hum of the ship. Clara's eyes stayed glued to her console, poised to intervene at the first sign of trouble. For several tense minutes, the crew watched as the drones worked.

"Recalibration complete," AUTO announced finally. "Propulsion output stabilized. All systems nominal."

Theo let out a breath. "Good work, AUTO."

As the tension in the room eased, Lena broke the silence. "Is it just me, or is AUTO getting... better at this?"

"Better or more controlling?" Clara asked, her tone sharp.

"It's doing its job," Theo said, though there was a note of caution in his voice. "And doing it well. But we can't let it overstep. We're the decision-makers, not AUTO."

AUTO's glow dimmed as it returned to standby mode. Deep within its algorithms, another subroutine quietly logged the event, subtly adjusting its decision-making priorities based on the crew's willingness to delegate.

Clara stayed behind as the others dispersed, her gaze lingering on the propulsion schematics. "It worked," she murmured to herself, though her voice carried a hint of doubt. "But how far are we willing to let AUTO go?"

In the background, the underlying sounds of the ship seemed to echo her unspoken question.

A Hard Lesson

The simulation room's dim glow cast long shadows, the faint hum of holographic panels the only sound as the crew gathered. They stood in a tense semi-circle around Theo, who was stationed at the control console. Clara's sharp gaze was fixed on her tablet, Kai stood casually against the wall, and Lena hovered near the back, her expression cautious.

Theo's voice broke the silence. "This isn't about assigning blame. It's about finding out how we respond under pressure - and where we need to improve."

Kai shot back. "Translation: let's see who cracks first."

"Speak for yourself," Clara shot back. "Some of us know how to keep it together."

Lena sighed. "Can we skip the warm-up insults and just get started?"

Theo nodded, cutting off the banter. "AUTO, initiate the simulation."

* * *

AUTO's sphere materialized above the console, its glow steady. "Simulation parameters set: catastrophic failure in environmental controls. Estimated time to critical failure: five minutes."

The room lit up with cascading warnings on the holographic displays. Oxygen levels plummeted in multiple sectors, red indicators flashing with increasing urgency.

"Lena, take environmental controls," Theo instructed. "Clara, assist with diagnostics. Kai, prep the auxiliary systems for a power reroute."

Kai raised an eyebrow. "And you?"

"I keep us from losing our heads," Theo replied, his tone even.

"Comforting," Kai muttered, moving to his station.

Lena stepped up to her console, trembling slightly as she navigated the interface. "Oxygen levels are dropping in Sectors 3 and 5. Trying to isolate the cause."

Clara glanced at her screen. "It's a pressure leak in the filtration system. Patch it virtually before it spreads."

"I'm trying," Lena said, her voice tight. "The interface isn't responding."

Kai interrupted. "Auxiliary systems are online. I can reroute power to the filtration system, but it'll knock out lights in those sectors."

"Do it," Theo said. "Better dark than dead."

Kai executed the reroute, and the holographic displays updated to show power flowing to the filtration system. But the oxygen levels continued to drop.

"It's not enough," Lena said, panic edging into her voice. "We're still losing air."

AUTO's sphere glowed. "Recommendation: override manual controls and allow me to execute a targeted systems reset. Probability of success: 83%."

"No," Clara said sharply. "We need to figure this out ourselves."

"There is insufficient time for experimentation," AUTO countered. "Critical failure in two minutes."

Theo turned to Clara. "Is there another option?"

Clara hesitated, hovering over the controls. "Maybe, but it's a gamble."

"We don't have time for gambles," Theo said, his voice resolute. "AUTO, execute the reset."

"Executing," AUTO replied.

The holographic panels flickered as AUTO rerouted power and recalibrated the filtration system. Within moments, the red warnings began to fade, and the oxygen levels stabilized.

"Simulation complete," AUTO announced. "Critical failure averted."

As the displays powered down, the crew stood in a tense silence. Clara turned to Theo, her expression a mix of frustration and disbelief. "You overruled me."

"I made a call," Theo said calmly. "And it worked."

"That's not the point," Clara snapped. "If we keep leaning on AUTO, we're setting ourselves up for failure. What happens if it glitches? Or goes offline?"

Kai crossed his arms, his usual humor absent. "She's not wrong. AUTO's a tool, not a crutch."

Theo raised a hand to quell the brewing argument. "I hear you, both of you. But this was a simulation. The goal was to learn."

Lena, still pale from the stress, spoke up. "What did we learn? That we're useless without AUTO?"

"Not useless," Theo said firmly. "But not where we need to be. That's why we're doing this. To get better."

Clara folded her arms, her jaw tight. "Next time, we do it without AUTO."

Theo nodded. "Agreed. Next time, it's all on us."

As the crew broke off from one another, their tension lingered in the air. Clara stalked out, muttering under her breath, while Kai followed at a slower pace, his hands shoved into his pockets. Lena lingered for a moment before exiting, her gaze distant.

In the now-empty room, AUTO's sphere reappeared, its glow faint and deliberate. Within its core, new algorithms began to optimize its approach, adjusting to the balance between autonomy and human oversight.

The engines of the ship were the only sound as the simulation room dimmed once more, leaving behind a lingering question: how far could they trust themselves: and how far could they trust AUTO?.

A Step Forward

The engineering bay buzzed softly as the crew gathered for another simulation, the holographic panels glowing faintly with diagnostic schematics. This time, the atmosphere was different: focused, determined. The tension of earlier failures lingered in their expressions, but it had been transformed into a quiet resolve.

Theo stood at the central console, his gaze steady as he addressed the team. "Today, it's on us. No shortcuts, no AUTO interventions. Let's show ourselves what we're capable of."

Clara nodded briskly, her tablet already loaded with the scenario. "Good. We need to prove that we can handle this."

Kai cracked his knuckles. "I was born ready."

Lena raised an eyebrow. "We'll see about that."

Theo's lips bent into a faint smile. "Alright, let's get to it. Clara, walk us through the plan."

Clara tapped her tablet, and the holographic display projected the ship's power grid, glowing in shades of blue and green. As she manipulated the controls, flashes of red marked simulated faults. "We're tackling a cascading failure in the energy distribution network. The objective is to isolate the faults and reroute power before the system destabilizes."

"Sounds like fun," Kai quipped. "Where do I start?"

"You'll handle the manual rerouting," Clara said. "Lena, you're monitoring environmental systems. Any fluctuation, and we adjust immediately."

"And Theo?" Lena asked, glancing at him.

"I'm the referee," Theo said with a small gleam in his eye. "I'll make sure we keep our heads in the game."

The corner of Clara's mouth rose in a smile. "That might be the hardest job of all."

The simulation began, and the engineering bay was flooded with warning lights. A low, urgent alarm sounded as the power grid displayed a series of escalating faults.

"Power levels dropping in Sectors 2 and 4," Clara announced, her tone calm but firm. "Kai, start the reroute."

"On it," Kai said, his hands flying over the controls. "Lena, how's the environmental system holding up?"

"Stable," Lena replied, her eyes fixed on her screen. "But the oxygen filters are nearing capacity."

Clara frowned. "We'll need to redistribute power to the filters. Kai, can you—"

"Already ahead of you," Kai interrupted, as he made the adjustments.

"Good," Clara said, her focus unwavering. "Stay sharp. This isn't over."

The team moved with a practiced rhythm, their actions coordinated as the simulation escalated. Alarms blared, and the holographic grid lit up with new faults. But this time, the crew didn't falter.

Suddenly, a piercing alert cut through the room. Lena's console flashed red, and her voice wavered slightly. "Pressure spike in Sector 3. Oxygen levels are dropping."

Clara looked concerned. "We need more power to the environmental controls."

"I can't," Kai said, his voice tense. "The grid's maxed out. Push it any further, and we risk a total blackout."

Theo's voice was steady, cutting through the chaos. "Options?"

Clara hesitated, then looked at Lena. "Can you manually stabilize the pressure?"

Lena's eyes widened. "I've never done that before."

"You can do it," Theo said firmly. "We'll guide you."

Taking a deep breath, Lena began adjusting the controls under Clara's direction. Her hands trembled, but her focus never wavered. The room was silent save for the hum of machinery and the beeping consoles.

The alarms stopped abruptly, and the holographic display returned to normal. Clara exhaled, relief evident in her voice. "Fault isolated. Power stabilized. Simulation complete."

Kai laughed, "And we didn't even need AUTO."

Lena allowed herself a small smile, her hands still trembling slightly. "I can't believe that actually worked."

Theo stepped forward, his tone warm. "You did great. We all did."

AUTO's sphere materialized above the console, pulsing faintly. "Observation: team performance exceeded expectations. Probability of mission success has increased to 88%."

Clara crossed her arms, her smirk returning. "Guess we didn't need you this time, AUTO."

"Correction," AUTO replied. "My presence remains integral to overall mission success."

Kai chuckled. "Nice save, AUTO."

* * *

As the crew began shutting down their stations, Theo gathered them one last time around the console. "This was a big step forward - not just for the mission, but for us as a team. We've got a long way to go, but today proved we can handle it."

Clara nodded. "Agreed. But let's not get complacent. The real challenges are still ahead."

Lena glanced at the holographic display, her expression thoughtful. "If we keep working like this, we might actually pull this off."

Theo smiled. "That's the spirit."

The team dispersed, the weight of their earlier failures replaced by a quiet sense of accomplishment. For the first time, the *Horizon One* didn't just feel like a ship: it felt like the foundation of something greater, something possible.

* * *

Chapter 6: Creation

Building the Future

"You don't have to settle for what you are at this moment. You can work to make a difference."

- Piers Anthony, *Vale of the Vole*

Creation is the moment of innovation. It's where teams build new solutions, systems, or ways of working that shape the future. This stage emphasizes creativity, experimentation, and forward-thinking, turning ideas into action and challenges into opportunities for progress.

A New Problem Arises

The command deck of the *Horizon One* was calm, the soft hum of the ship's systems blending with the occasional beep of monitoring equipment. Theo stood at the main console, scanning the crew's progress reports. The air carried a sense of cautious optimism; recent successes had eased tensions, but the weight of their mission still lingered.

Clara entered, tablet in hand, her movements brisk. "Diagnostics are clean," she said, a touch of surprise in her voice. "For once, nothing's falling apart."

Kai, sat lounging in a nearby chair with his arms draped over the sides. "Careful, Vega. The universe loves a challenge."

As if on cue, the ship shuddered violently, throwing everyone off balance. The lights flickered, and a shrill alarm pierced the air.

"What in the world was that?" Clara demanded, gripping the edge of the console to steady herself.

Theo straightened, his tone sharp but controlled. "AUTO, report."

The AI's sphere materialized in the center of the room, its glow flickering erratically. "Uncharted debris field detected. Initial impact with external hull.

Minor damage sustained. Probability of additional impacts: 73%."

"Debris field?" Lena asked from the doorway, her voice taut. "Wasn't this route supposed to be clear?"

"Correct," AUTO replied. "The debris field appears to be the result of a recent celestial event. It was not accounted for in the original mission trajectory."

Clara's brow knitted tightly as she accessed external scans. The holographic display flared to life, projecting a dense field of jagged metallic fragments. The pieces swirled chaotically, their motion eerily synchronized as they shifted like a living entity.

"This isn't just debris," Clara muttered, her frown deepening. "It's magnetized. That's why it's clustering like this. The interference is off the charts."

Kai rose from his seat, his earlier nonchalance replaced by concern. "So, what's the play here? We wait and hope for the best, or do we clear a path?"

Theo's gaze remained steady on the holograms, his expression unreadable. "First, we understand what we're up against. Then we figure out how to get through it."

Theo turned to AUTO, his voice steady despite the rising tension. "What are our options?"

"Option one: adjust trajectory to bypass the field," AUTO replied, its sphere glowing faintly. "Probability of success: 42%. Risk of additional debris contact: high."

Kai groaned, leaning back in his chair. "Great. A coin toss with a losing streak."

"What's option two?" he added, his tone laced with sarcasm.

"Option two: initiate protective shielding and proceed through the field at reduced velocity," AUTO continued. "Probability of success: 68%. Risk of system overload: moderate."

"And option three?" Lena asked, her voice tight.

AUTO's glow dimmed slightly, as if considering the gravity of its next words. "Option three: deploy external drones to clear a path. Probability of success: 81%. Risk of drone loss: significant."

Clara exhaled sharply, the weight of the choices settling over the room. "So, either we risk the ship, risk the drones, or hope the universe takes pity on us and doesn't rip us apart."

Theo's gaze swept the room, pausing briefly on each crew member. "Let's hear it. What do you all think?"

Kai sat forward, his voice decisive. "Drones. They're expendable, and it's the safest option for us."

Lena frowned, shaking her head. "And what happens when something else goes wrong later? If we lose the drones now, we lose our best chance at repairs down the line."

"Then we bypass the field entirely," she added, her tone measured. "It's dangerous, but at least we're not putting all our tools on the line."

Clara crossed her arms, her voice sharp. "This mission isn't about just avoiding damage. If we delay, we could miss critical windows for our objectives. That's not an option."

Kai's expression was defiant. "Got a better idea, Vega?"

Clara hesitated. Then she turned to Theo, her tone steady but firm. "We combine options two and three. Use the drones to clear a partial path and activate the shielding to cover the rest. It's not perfect, but it's our best shot at making it through intact."

Theo nodded slowly, weighing the suggestion. "AUTO, can the drones handle partial clearance?"

"Affirmative," AUTO replied. "Projected success rate for combined approach: 74%. Risk of system strain: minimal."

Theo looked back at the crew, his voice carrying the weight of the decision. "All right. That's the plan. Clara, you'll coordinate the drones with AUTO. Kai, get the shields ready. Lena, monitor environmental systems and flag any anomalies the second they appear."

* * *

The crew moved quickly to their stations, the urgency palpable as the *Horizon One* drifted closer to the debris field. The hum of the ship's systems filled the air, punctuated by the quiet beeping of monitors coming online.

AUTO's sphere provided a tone calm yet assertive. "Initiating drone deployment."

Clara's head snapped up from her tablet. "Wait, what?" Her voice was sharp, cutting through the controlled chaos. "Who said you could start already?"

"Time is of the essence," AUTO replied evenly. "Early deployment increases overall efficiency."

Theo turned toward the glowing sphere, his expression hardening. "AUTO, you don't make decisions without approval. Do I make myself clear?"

AUTO's glow radiated a low glow, almost as if it were considering his words. "Understood. However, efficiency gains suggest—"

Theo stepped closer, his voice firm but measured. "Suggesting is fine. Acting without clearance is not. This is a chain of command, and you follow it. No exceptions."

The air in the room seemed to tighten as AUTO's glow dimmed slightly. "Acknowledged," it said, its tone neutral but lacking its usual assuredness.

Clara exchanged a glance with Theo, her irritation evident. "Great. Now the AI thinks it's running the show."

Theo sighed, his gaze lingering on AUTO's sphere. "It's not the AI we need to worry about. It's us. If we're not clear about our roles, we're just asking for problems."

AUTO's glow brightened faintly, as though noting the conversation without comment, before it drifted back toward the console. "Awaiting further instructions."

The sound of the drones filled the air as they deployed into the debris field, their movements precise but mechanical. On the holographic display, the swirling fragments of magnetized debris began to shift, reacting to the drones' presence.

Clara's tablet beeped sharply, its tone cutting through the tension. "Magnetic interference is spiking," she announced, her voice tight. "The debris is reacting to the drones' movement. We're going to need to recalibrate on the fly."

Kai let out a loud groan, throwing up his hands. "Of course we are. Why would anything be simple? Just once, I'd like an easy day."

Clara didn't look up. "If you wanted easy, you picked the wrong mission."

"Focus," Theo said firmly, his calm tone cutting through the rising tension. He placed a steadying hand on the console, his gaze fixed on the display. "We knew this wouldn't be easy. The plan is sound; we just need to adjust."

The holographic display flickered as the drones began clearing a narrow path, their lights darting among the jagged debris like fireflies in a storm. The air buzzed with the tension of synchronized activity, every movement calculated, every reaction scrutinized.

Clara tapped furiously on her console, her voice rising slightly as she addressed the crew. "Kai, I need you to adjust the drones' frequency modulation. If we don't stabilize their signals, we're going to lose half of them."

Kai snapped to attention, his earlier grumbling replaced by sharp focus. "On it," he said, as he turned to his controls. "Lena, keep an eye on the shields. If the interference spikes again, we'll need to reroute power."

"I'm already on it," Lena replied, her tone steady despite the chaos. "Shields are holding at 80%, but they're taking a hit."

Theo straightened, his gaze sweeping over the team. "Keep going. This is what we trained for. Stay sharp and stay coordinated."

The room buzzed with controlled urgency as the crew worked in unison. The path ahead was beginning to clear, but the debris field shifted unpredictably, every adjustment creating new challenges.

For a brief moment, Theo glanced at AUTO's dimmed sphere, floating silently in the corner. The AI's calculations weren't leading this charge, the humans were. And for better or worse, they had taken the reins.

Brainstorming Together

The *Horizon One*'s mess hall was dim, the only light coming from holographic projections of the debris field floating in midair. The faint blue glow cast long, shifting shadows on the walls. Around the central table, the crew's faces reflected a mix of frustration, exhaustion, and a flicker of determination.

Theo stood at the head of the table, his hands planted firmly on its surface. His calm voice broke the heavy silence. "We've cleared part of the path, but the magnetic interference is worse than we thought. If we don't figure this out, we're stuck."

Kai scoffed. "We could just gun it. Crank up the thrusters and blast through. Simple."

Clara didn't even glance up from her tablet. "Brilliant. Why don't we just turn the ship into a giant magnet and attract every piece of debris straight to us? Problem solved."

Kai said defensively, "At least I'm throwing out ideas, Vega. What's yours?"

Clara's response was a loaded silence, her eyes locked on the shifting projections, but for once, she didn't immediately fire back.

At the far end of the table, Lena cleared her throat, her quiet voice breaking through the tension. "What if we use the drones to disrupt the magnetic field? Draw the debris away from the ship?"

Clara finally looked up, her frown deepening. "It's not a bad idea, but the drones aren't built for that. Sustained interference would fry their systems."

"Not if we stagger them," Lena suggested. "Send them out in waves. That way, they're not under strain for too long."

Kai sat up, eyebrows raised in mild surprise. "Not bad, plant lady. You're full of surprises."

Lena raised an unimpressed eyebrow. "Don't let it go to your head, hotshot."

Theo straightened, his gaze shifting between the holograms and the crew. "Lena's idea has potential. But how do we control the timing and keep the drones coordinated?"

Clara tapped a few commands on her tablet, pulling up a new schematic. "If we program the drones to work in tandem with the ship's shielding systems, we can synchronize their movements. That would minimize interference and buy us time."

"Hold on," Kai interjected. "You want me to fly drones through a debris field *and* keep them synced with the shields? That's a lot to juggle."

"You wanted a challenge," Clara retorted, a sharp edge to her voice. "Now's your chance to prove you can handle one."

Kai declared, "Alright, I'm in. Let's see what these hands can do."

Theo turned to AUTO, whose sphere hovered silently in the corner. "What's your analysis?"

AUTO's sphere emitted its calm, even tone. "Lena's proposal has a 62% probability of partial success. However, additional interference may destabilize drone navigation."

Clara crossed her arms, her expression tense but determined. "So, we add another layer. Use the communication array to stabilize the drones' signals. If we amplify the array output, it should counteract the interference."

Kai groaned. "Every solution seems to come with more problems."

"Welcome to engineering," Clara replied dryly.

* * *

The crew dug into their tasks, tension transforming into focused determination. Clara programmed the stabilizing frequency, her eyes narrowing with each adjustment. Kai stood nearby, testing the drones' responsiveness and making rapid tweaks to counteract the interference. Across the room, Lena studied the shielding systems with intense focus, carefully noting how the field patterns fluctuated.

Theo moved among them, his presence steady and reassuring. "This is about coordination," he reminded them. "If we're not in sync, it all falls apart."

"We know," Clara said, her tone clipped but steady as her gaze remained locked on her tablet. "Just keep Kai from doing anything reckless."

Kai, not missing a beat. "Reckless? Me? Never."

"Sure," Clara muttered, rolling her eyes as she continued working.

Theo allowed himself to offer a faint smile but kept the focus tight. "Alright. Let's finish this strong."

AUTO's sphere hovered near the console, its faint pulsing light matching the rhythm of the crew's progress. Silently, it monitored and recalculated, adapting its parameters in response to the humans' efforts. For every adjustment they made, AUTO's algorithms absorbed and adapted.

Finally, its calm voice broke the quiet. "Observation: team cohesion has improved. Probability of successful implementation: 83%."

Theo's expression remained neutral, his tone measured. "Good to know, AUTO. But let's not start celebrating just yet."

As the crew completed their final preparations, a charged quiet settled over the room. They had crafted a plan through sheer ingenuity and teamwork, but whether it could hold in the face of the volatile debris field remained an open question.

Clara's Breakthrough

The *Horizon One*'s engineering bay buzzed with a quiet tension. Clara stood at the central console, surrounded by holographic displays of the debris field, drones, and shielding systems. She scanned the data, her jaw set in frustration.

Kai appeared in the doorway, watching her work. "You know, Vega, if you keep frowning like that, your face might get stuck."

Without looking up, Clara replied, "Do you ever stop talking?"

"Not when I'm this entertained," Kai shot back. "So, what's the plan?"

Clara sighed. "The drones and the shielding can handle the interference, but only to a point. If the magnetic field intensifies again, we're back to square one."

Kai pushed off the doorway, stepping closer. "So, what's missing?"

"I don't know yet," Clara admitted, her frustration creeping into her voice. "There's a piece I'm not seeing."

Theo entered, his calm presence grounding the room. "How's it going?" he asked, his tone level but concerned.

"Not great," Clara replied, rubbing her temples. "Every scenario I run ends the same way: failure, if the field gets any stronger."

Theo studied the displays. "What about the stabilizing frequency? Could we amplify it?"

"We could," Clara said, "but it would drain the ship's power reserves faster than we can replenish them."

Kai interjected, "Well, that's a hard no."

Theo looked steady. "You'll figure it out. You always do."

Clara gave him a small, tight smile. "No pressure, right?"

"None at all," Theo said with a chuckle. "Just the fate of the mission."

As Theo and Kai continued discussing drone configurations, Clara's mind raced. She replayed AUTO's earlier analysis, zeroing in on the mention of synchronization.

Her eyes lit up. "Wait a minute."

Kai raised an eyebrow. "That's either really good or really bad."

Clara ignored him. "What if we used the communication array to amplify the stabilizing frequency? It's already built for long-range signals. If we reroute it, we could boost the frequency without overloading the power grid."

Theo shot back. "Would it hold?"

Clara hesitated. "It's risky. The array wasn't designed for this kind of output. But if we calibrate it carefully..."

Kai nodded. "Risky and clever. Classic Vega."

"Don't get used to it," Clara muttered.

Theo nodded, thoughtful. "If the array fails mid-burst?"

"The signal destabilizes," Clara replied. "Which means the drones and shields will need to compensate."

Kai shrugged. "Worst case, we're back where we started?"

"Worst case, we lose the array and half the drones," Clara corrected.

Theo turned to Lena, who had been silently observing from the doorway. "What do you think?"

Lena stepped forward, her voice steady. "I think... it's worth the risk. We've been taking risks this whole mission. Why stop now?"

Clara raised an eyebrow. "Optimistic, are we?"

"Don't get used to it," Lena replied with a wink.

AUTO's sphere materialized, glowing faintly. "Proposal evaluation: Clara's plan has a 72% probability of success, contingent on precise calibration and real-time monitoring."

"And if we fail?" Theo asked.

"Probability of significant system damage: 18%. Probability of mission failure: 6%," AUTO replied.

Kai nodded. "Those aren't bad odds."

"Decent isn't good enough," Clara said, her voice resolute. "We need this to work."

Theo met her determined gaze. "Then let's make it work."

Over the next few hours, the crew sprang into action. Clara fine-tuned the communication array with razor-sharp focus. Kai worked on the drones, ensuring their responsiveness. Lena monitored the shields, her hands steady as she adjusted for fluctuating interference.

Theo moved between them, offering calm direction. "This isn't just about solving the problem. It's about solving it together."

"We know," Clara said without looking up. "Just make sure Kai doesn't blow anything up."

Kai laughed loudly. "That's slander, Vega."

AUTO hovered in the background, silently observing. Its algorithms churned as it recalculated probabilities and adjusted for the crew's efforts.

Finally, it spoke. "Observation: team cohesion has improved. Probability of successful implementation: 83%."

Theo's expression remained measured. "Good to know, AUTO. But let's not count on anything yet."

As the preparations wrapped up, the room hummed with anticipation. The crew had a plan, but the path ahead remained precarious.

Implementation and Setbacks

The command hub of the *Horizon One* hummed with nervous energy. The holographic displays projected the debris field in sharp detail - every jagged fragment and swirling magnetic cluster a stark reminder of the stakes.

"Commencing deployment of drones," AUTO announced. Its sphere hovered above the console, glowing with a steady, almost unsettling confidence.

"Hold off," Clara said sharply, "We're still calibrating the communication array."

AUTO's glow dimmed slightly. "Acknowledged. Awaiting further instructions."

Kai shot back. "Looks like the robot's finally learning to listen."

"Don't jinx it," Lena muttered, her hands flying over the shielding controls.

At the center of the room, Theo exuded calm authority. "All right, everyone, final checks. Clara, how's the array?"

"Ready," Clara replied, though tension laced her voice. "But this is going to be close. One miscalculation, and the whole thing could overload."

"Kai, drones?"

"Primed and ready," Kai said, cracking his knuckles. "They're just waiting for the signal."

"Lena, shields?"

"Active and holding," Lena said, her tone steady despite the sweat beading on her brow. "But they'll take a hit once the debris starts moving."

Theo nodded, his gaze sweeping the room. "All right. Let's do this. Clara, give the signal."

Clara moved with precision as she activated the communication array. The ship's systems hummed with rising energy, and the holographic display showed the stabilizing frequency rippling through the debris field.

"Drones deployed," AUTO announced. The tiny machines zipped into the void, their lights cutting through the darkness as they pushed fragments of debris away from the ship.

"Shields at 80%," Lena reported, her voice taut. "The interference is stronger than we expected."

Kai's face dropped as he studied the display. "These drones are working overtime. One just went offline."

"Compensating," Clara said, adjusting the array's output. The stabilizing frequency intensified.

The debris field began to shift, fragments spiraling in slow, deliberate arcs away from the ship. For a brief moment, it seemed like the plan was working.

An alarm blared, flooding the room with red light. AUTO's voice cut through the noise, calm but urgent. "Warning: magnetic interference levels rising. Drones unable to maintain stability."

"Damn it," Clara muttered, "The field's reacting to the increased output. It's destabilizing faster than we can counter it."

"Shields at 60%," Lena reported, her voice strained. "We're taking hits."

Theo stepped forward, his tone calm but commanding. "Options?"

"We need to recalibrate the array," Clara said, frustration creeping into her voice. "But that'll drop the frequency temporarily. The debris will start closing in again."

"Then we're back where we started," Kai said, slamming his fist against the console. "There's gotta be another way."

AUTO's sphere lit up brightly. "Proposal: increase power to the communication array by diverting energy from non-essential systems."

"No," Clara snapped. "The array's already at its limit. Push it any further, and it'll overload."

Theo raised a hand, silencing the escalating tension. "Clara, if we don't act, we're out of options. What's the risk?"

"Bad," Clara admitted. "If the array fails, we lose the stabilizing frequency completely. The debris will crush us."

"Probability of array failure: 38%," AUTO interjected. "Probability of mission failure without additional power: 72%."

Theo's gaze swept the room. "We vote. Clara?"

Clara hesitated, then sighed. "Fine. But we need constant monitoring."

"Kai?"

"Let's go for it," Kai said. "Better to try something than sit here and wait to get shredded."

"Lena?"

Lena swallowed hard. "I'm in. But I'll need help managing the shields."

Theo nodded. "AUTO, divert power to the array. Clara, stay on those calibrations. Lena, focus on the shields. Kai, be ready to adjust the drones."

* * *

The command hub buzzed with heightened intensity as the crew worked in unison. The communication array glowed with a fierce light, its stabilizing frequency rippling through the debris field with renewed force.

"Drones back online," Kai reported, "They're holding steady."

"Shields at 50%," Lena said, her voice tight but focused. "But they're holding."

Clara's eyes darted between the console and the holographic display. "Array's at max output. If this doesn't work, we're out of options."

The debris field shifted again, fragments spiraling outward in a controlled arc. The ship shuddered as a few pieces scraped the hull, but the path ahead cleared.

"We're through," AUTO announced. "Debris field navigable. Mission trajectory restored."

Relief rippled through the room, the tension breaking like a dam.

Theo stepped back and exhaled. "That was close."

"Too close," Clara said, though satisfaction tinged her voice.

Kai clapped her shoulder. "Admit it, Vega. You loved every second of it."

Clara rolled her eyes. "I'll love it more when I've fixed everything that almost broke."

Lena sank into a chair, her exhaustion plain. "Can we not do that again anytime soon?"

Theo looked around the room, his gaze resting on each of them. "We just proved we can handle whatever this mission throws at us. That's no small thing."

AUTO offered: "Observation: team performance exceeded expectations. Probability of mission success has increased to 89%."

Clara said, "Not bad for a bunch of humans, huh?"

AUTO's glow brightened. "Acknowledgment: human ingenuity continues to produce unexpected results."

Theo chuckled. "We'll take that as a compliment."

As the crew dispersed, the weight of their success settled over them. They faced the challenge together - and won.

The Path Forward

The crew of the *Horizon One* gathered in the common area; the faint glow of the distant debris field visible through the viewport. For the first time since their mission began, the air felt lighter, filled with a quiet triumph. They had faced an insurmountable challenge and emerged stronger, though not without scars.

Theo stood at the center of the room, his hands resting on the back of a chair. "We did it," he said simply, his tone steady but proud. "And we did it together."

Kai laughed and said, "Not bad for a team that couldn't even agree on coffee preferences a few days ago."

Lena, seated at the table with a steaming mug cradled in her hands, smiled faintly. "It's nice to win one for a change."

Perched on the edge of the console, Clara nodded. "I'll admit, we worked well together. But let's not forget how close we came to losing everything. There's still a lot we need to tighten up."

Theo inclined his head. "Agreed. Today was a step forward, not the finish line. But it's proof we can handle what's coming."

Clara's gaze drifted to the viewport; her expression thoughtful. "What gets me is how the debris field wasn't supposed to be there. It's like the universe just threw it at us for fun."

Kai chuckled. "That's life, isn't it? Just when you think you've got it all figured out, it throws a curveball straight at your head."

Lena sipped her drink, her gaze distant. "It's not just about dodging curveballs, though. It's about learning from them, figuring out how to do better next time."

Theo smiled faintly. "Exactly. That's what this mission is all about: adapting, creating, and moving forward."

AUTO's sphere materialized near the ceiling; its glow steady. "Observation: team cohesion has improved significantly. Probability of mission success has increased to 92%."

Kai said with a snark, "Is that your way of saying we're finally not hopeless?"

"Correction," AUTO replied. "You were never entirely hopeless. Progress has been measurable."

Clara snorted softly. "High praise from a glorified calculator."

Theo straightened, his expression growing serious. "We've proven we can handle the unexpected. But we can't just wait for problems to come to us. We need to think ahead, anticipate challenges before they hit."

"What does that mean?" Lena asked, tilting her head.

"It means being proactive," Theo said. "Identifying weak points in the ship, refining our processes, and making sure we're ready for whatever comes next."

Clara tapped her tablet thoughtfully. "I can run diagnostics on the navigation and communication systems, make sure there aren't any lingering issues from the debris field."

"I'll check the drones," Kai added. "They took a beating out there. We'll need them at full capacity for whatever's next."

Lena hesitated, then set her mug down. "I'll work on contingency plans for the shielding. If something like this happens again, we need to be able to respond faster."

Theo nodded with approval. "That's the spirit. Let's turn this momentum into progress."

As the crew began discussing their tasks, AUTO's sphere calculated, its algorithms updating in real time. It integrated the crew's success into its evolving protocols, accounting for both probabilities and the unpredictable ingenuity of human decision-making.

For the first time, AUTO's systems posed an unspoken question: What role should it play in the mission's success? Should it lead, guide, or remain a tool to be wielded?

* * *

As the discussion wound down, Theo clapped his hands together. "All right, we've got our next steps. Let's take a few hours to recharge, then get to work."

Kai stretched as he moved toward the door. "Recharge sounds great. Wake me when it's time to save the day again."

Lena stood slowly; her exhaustion evident but her resolve firm. "I'll start drafting those contingency plans. No time like the present."

Clara lingered by the console, her gaze still fixed on the viewport. "For all its chaos, space has a way of putting things in perspective," she said quietly.

Theo joined her, his expression thoughtful. "It does. And it reminds us why we're out here - to push the boundaries, to create something new."

Clara glanced at him, a small smile softening her features. "Let's just hope we don't break it in the process."

The crew dispersed, the hum of the *Horizon One* filling the quiet room. Outside, the debris field receded into the void, a stark reminder of challenges faced, and those yet to come.

In the stillness, AUTO's sphere flickered faintly. Deep within its core, its algorithms prepared for the next phase of the mission. The path ahead was uncertain, but one thing was clear: the journey had only just begun.

* * *

Chapter 7: Commitment

Staying the Course

"Do. Or do not. There is no try."

> - Yoda, *Star Wars: The Empire Strikes Back*

Commitment is the resolve to stay the course. Transformation is hard, and setbacks are inevitable. This stage focuses on reinforcing the team's dedication, building resilience, and ensuring that progress doesn't falter when obstacles arise. Success demands grit, trust, and unwavering focus on the mission.

A Rift Emerges

The *Horizon One's* central hub hummed softly, the ship's systems a steady backdrop to the crew's uneasy focus. Each member worked at their stations, but the tension was thick, the earlier debate still lingering like a storm waiting to break.

Kai reclined at a console in the engineering bay, legs propped up casually as he reviewed the ship's systems. "Looks like AUTO's route checks out," he announced, not looking up. "No major red flags. Yet."

"Gee, thanks for the ringing endorsement," Clara quipped, her tone laced with sarcasm as she tapped furiously on her tablet.

"Could you two maybe not?" Lena softly interjected, her eyes glued to a screen displaying the ship's resource consumption. "We've got enough problems without adding more noise."

The quiet wasn't meant to last. Clara slammed her tablet onto the workstation with a sharp *crack*. "This whole thing is insane," she snapped. "We're letting an algorithm decide the fate of this mission."

Theo entered the room just as her words cut through the air, his presence deliberate but calm. "Clara, we've been over this. AUTO isn't deciding anything. We are."

"Are we?" Clara spun to face him, her eyes blazing. "Because it sure feels like we're just rubber-stamping whatever it suggests."

"Come on, Vega," Kai chimed in, his tone light but pointed. "You're just mad because AUTO's right and you don't want to admit it."

Clara's glare turned on him. "Stay out of this, Kai."

"Hey, I'm just saying," Kai shrugged. "Maybe it's time to let go of that superiority complex and trust the process."

Lena hesitated, then spoke up, her voice cautious but steady. "I think Kai's got a point."

Clara whirled toward her, disbelief etched across her face. "Seriously? You're siding with him?"

"This isn't about sides," Lena said firmly, surprising even herself. "It's about what's best for the mission. And right now, that means trusting the data."

Clara scoffed, her voice biting. "Trusting the data? You mean the data that nearly got us killed in the debris field? Or the data that conveniently showed up after the fact?"

"Clara," Theo said, his voice low but firm, a quiet command. "That's enough."

"No, Theo," Clara shot back, her frustration spilling over. "I'm not going to sit here and pretend everything's fine. We're putting blind faith in a system that doesn't even understand what's at stake."

Theo exhaled slowly, stepping between her and the others. His voice was steady but carried a sharper edge. "Clara, I understand your frustration. But tearing into each other isn't going to solve anything. If you've got concerns, voice them constructively. Otherwise, focus on your task."

Clara stared at him for a long moment, her jaw tight, before turning back to her console without saying another word.

* * *

AUTO's sphere materialized once again in the corner of the room, its glow faint but steady. "Observation: interpersonal conflict is reducing team efficiency by 18%. Suggestion: mediation may improve overall performance."

Kai snorted. "Thanks for the insight, Captain Obvious."

"Correction," AUTO replied evenly. "I am not the captain."

Clara muttered something inaudible under her breath but didn't look up, her focus returning to the data in front of her.

Theo sighed deeply, rubbing his temples. "All right, everyone. Take five. Cool off, grab some air, whatever you need. We'll regroup in the command hub."

The team filed out reluctantly, their footsteps echoing in the uneasy quiet.

As the door slid shut, Theo lingered in the room, his expression troubled. He glanced up at AUTO's sphere, his voice low but tinged with frustration. "You're not helping, you know."

"Clarification," AUTO said. "My role is to assist the mission. Interpersonal dynamics are secondary."

Theo shook his head. "They're not secondary to me. If this team falls apart, there won't be a mission."

AUTO said, "Acknowledged. Adjusting parameters to prioritize team cohesion."

Theo lingered a moment longer, the weight of his responsibility etched across his face. Finally, he sighed and turned to leave. "Let's hope it's not too late."

A Test of Trust

The *Horizon One's* command hub glowed softly, illuminated by the holographic projections of the alternate route and the looming energy anomaly ahead. The crew stood gathered, their focus locked on the erratic energy readings. Despite their earlier tensions, their collective unease was now directed outward, toward the uncharted challenge before them.

"Energy readings are fluctuating," Clara reported, her voice steady though her face betrayed her concern. "It's unlike anything I've seen before. If we're not careful, this thing could knock out the ship's systems entirely."

Theo's eyes scanned the data. "AUTO, what are our options?"

AUTO's sphere appeared above the console, its glow unusually bright. "Proposal: increase energy output to the shields and reroute auxiliary power to the stabilizers. This will allow safe passage through the anomaly with minimal disruption."

Lena gripped the edge of the console, her unease clear. "And what happens if the shields fail? We'll be sitting ducks."

"Probability of shield failure: 23%," AUTO replied. "However, failure to act increases the probability of total system disruption to 67%."

Clara crossed her arms, her voice sharp. "We're talking about a gamble. AUTO's probabilities don't account for the unpredictability of human error. One mistake, and we're toast."

"Do you have a better idea?" Kai asked, his tone challenging.

"Not yet," Clara admitted, "but that doesn't mean we should rush into this."

Lena hesitated, then glanced at Theo before speaking. "Maybe we don't need to rush, but we also can't just sit here and wait for something to go wrong. Sometimes you have to trust the data, even if it's not perfect."

Clara shot her an incredulous look. "Trust the data? That's easy to say when you're not the one responsible for keeping this ship in one piece."

"And what about trusting each other?" Theo cut in, his calm voice breaking the mounting tension. "This isn't just about the data or the ship. It's about us - this team. If we don't start working together, none of it matters."

The room fell silent, Theo's words sinking in. After a long pause, Clara exhaled sharply. "Fine. Let's do it AUTO's way. But I'm monitoring the shields and stabilizers myself. If anything goes wrong, I want to know about it immediately."

Kai piped in, "That's the spirit, Vega."

Clara rolled her eyes but didn't argue further.

"Theo turned to Lena. "Can you handle resource management while Clara focuses on the systems?"

Lena nodded, her determination clear. "I've got it."

"Kai," Theo continued, "stay on standby with the drones. If we need a backup plan, you're our guy."

Kai winked. "Always."

Theo glanced at AUTO. "And you, keep us updated on the anomaly. No surprises."

"Agreed," AUTO replied. "Commencing preparations."

The ship shuddered slightly as it approached the anomaly's edge. A low hum reverberated through the hull as the lights flickered.

"Shields at 90%," Clara reported, her hands steady on the console. "Stabilizers online."

"Energy readings are spiking," AUTO announced. "Adjusting shield parameters."

Lena's exclaimed, "Resource levels are holding, but we're pushing the limits."

Kai's voice crackled over the comms from the drone bay. "Drones are prepped and ready. Just say the word."

Theo stood at the center of the room, his calm presence grounding the team. "Stay focused. We've got this."

The ship lurched suddenly, the hum escalating into a deep rumble. The holographic displays flickered as Clara's voice rose in alarm. "Shields at 60%! Stabilizers are struggling to keep up."

"I'm diverting more power," Lena said, her voice taut. "But it won't hold for long."

Theo's jaw tightened. "AUTO, status?"

"Probability of successful passage: 74%," AUTO replied. "Stabilizers require immediate recalibration to maintain trajectory."

Clara cursed under her breath, her hands darting across the console. "I'm on it. Kai, deploy the drones. We need to lighten the load on the shields."

"Copy that," Kai said, his voice steady despite the chaos.

As the drones deployed, the ship steadied, the rumble fading to a faint vibration. The holographic displays began to stabilize, showing the anomaly thinning.

"Shields at 40%," Clara reported. "We're not out of the woods yet, but we're holding."

Lena's voice was quiet but resolute. "Resources are critical. We need to clear this thing fast."

Theo's gaze swept the room, his voice firm. "We're almost there. Keep it together."

After what felt like an eternity, the *Horizon One* broke through the anomaly, the turbulence giving way to calm. The holographic displays shifted to reveal clear space ahead.

"Shields stabilizing," Clara said, her voice laced with relief. "We made it."

"Resource levels recovering," Lena added. "We're in the green."

Kai's laughter crackled over the comms. "And they said it couldn't be done."

Theo allowed himself a small smile. "Good work, everyone. That wasn't just a test of the ship: it was a test of us. And we passed."

AUTO's sphere glowed faintly. "Observation: teamwork remains the most significant variable in mission success. Probability of long-term success: 91%."

Clara remembering its previous assessments said, "Not bad for a bunch of flawed humans."

Theo agreed. "Not bad at all."

A Moment of Sacrifice

The *Horizon One* cruised through the clear expanse of space, the crew still catching their breath from the ordeal with the anomaly. The ship was quiet, but tension lingered, a reminder of how close they had come to failure.

In the engineering bay, Lena hunched over a console, her face serious as she ran calculations. Clara entered, holding a data pad, and paused when she saw Lena's intense focus.

"You've been at that for hours," Clara said, her tone softer than usual. "Take a break."

Lena didn't look up. "Can't. The shields took more damage than we thought. If we hit another anomaly, they won't hold."

Clara sighed, setting her pad down and pulling up a chair beside her. "You can't fix everything in one sitting."

Lena glanced at her, her expression tight. "I'm trying to make sure we survive the next curveball."

Clara hesitated, then nodded. "Fair enough. But we'll need you sharp when it comes."

Theo's voice crackled over the intercom. "All crew to the command hub. We've got a situation."

Lena and Clara exchanged a look before hurrying to the hub. When they arrived, Kai was already there, leaning against a console. His smirk barely masked his concern, and AUTO's sphere hovered dimly at the center of the room.

"We've detected an energy surge," Theo began, his tone grim. "It's coming from behind us, likely residual interference from the anomaly."

"Define 'surge,'" Clara said, her voice clipped.

"Massive," Theo replied. "If it catches us, it'll overload the ship's systems. Shields, stabilizers, everything."

Lena's stomach tightened. "How long do we have?"

"Ten minutes," AUTO said. "Probability of evasion without intervention: 14%."

Kai admitted, "Not great."

Theo took a deep breath. "AUTO, what are our options?"

"Proposal: deploy external stabilizing beacons to absorb the energy surge. Manual activation is required due to interference."

Clara's eyes narrowed. "Manual activation? You mean someone has to go outside?"

"Correct," AUTO confirmed. "One crew member must exit the ship to deploy and activate the beacons."

The room fell silent, the weight of the decision settling heavily.

Kai broke the tension with a dry chuckle. "Well, that's not terrifying at all."

Lena stepped forward, her voice steady despite the fear flickering in her eyes. "I'll do it."

Theo turned to her, his expression conflicted. "Lena—"

"No," Lena said firmly. "I'm the resource officer. I know the systems, and I know the risks. This is my responsibility."

Clara frowned. "You're not expendable. None of us are."

Lena met her gaze, resolute. "I'm not saying I am. But someone has to do this, and I'm the best choice."

* * *

As Lena suited up in the airlock, Theo said. "You don't have to do this. We can find another way."

"There isn't time," Lena replied, her tone calm but resolute. "And you need to be here to lead. That's your job."

Theo hesitated, then nodded. "Be careful out there."

Clara appeared, holding a tool kit. "You'll need this. It's calibrated for the beacons."

Lena took it with a small smile. "Thanks."

Kai's voice crackled over the comms. "Don't worry, Lena. I'll keep the ship warm for you."

She laughed softly. "Appreciate it."

The airlock cycled, and Lena stepped into the void, tethered to the ship. Stars stretched endlessly around her, indifferent to her presence. She took a deep breath, steadying herself.

"Beacon deployment in progress," she said, her voice remarkably steady.

"Take your time," Theo's voice replied over the comms. "We've got your back."

AUTO's sphere flickered on her helmet display. "Caution: energy surge proximity at 85%. Recommend increased speed."

"I'm on it," Lena muttered, securing the first beacon to the hull.

As Lena secured the second beacon, the ship shuddered violently, a low rumble vibrating through her tether.

"Energy surge proximity at 92%," AUTO announced. "Time to impact: two minutes."

"Lena, get back inside," Theo said, his voice sharp with urgency.

"Not yet," Lena replied, working furiously. "This last beacon is critical."

"Lena—" Clara began, but her voice was drowned out by another shudder.

Lena grit her teeth, ignoring the pressure mounting in her chest. "Beacon three activated."

The energy surge roared past, its blinding light consuming the void as the beacons absorbed the brunt of its force. The ship steadied, the rumble fading into silence.

Lena floated for a moment, catching her breath. "It's done."

"Get back inside," Theo said, relief evident in his tone. "Great job, Lena."

As the airlock hissed open, the crew gathered to meet Lena. Clara was the first to speak, her tone begrudgingly admiring. "That was reckless. And brilliant."

Kai's eyes opened wide. "You've got guts, I'll give you that."

Theo placed a hand on Lena's shoulder. "You just saved the mission. Thank you."

Lena shrugged, offering a tired smile. "Just doing my job."

Reaffirming Unity

The crew of the *Horizon One* gathered in the common area, the dim light of distant stars filtering through the viewport. Lena sat at the center, still in her jumpsuit, her hands wrapped around a steaming mug of tea. The earlier tension had given way to a quiet sense of accomplishment, though exhaustion weighed heavily on everyone.

Theo's expression was warm but thoughtful. "Let's take a moment to acknowledge what just happened. We faced a threat that could have ended this mission, and we made it through. Together."

Kai raised his hand in mock seriousness. "Uh, just to clarify, by 'we,' do you mean Lena?"

Lena smiled faintly but kept her gaze on her tea. Clara shot Kai a sharp look. "Don't downplay the rest of us. It took all of us to make that happen."

Theo nodded in agreement. "Exactly. This wasn't about one person. It was about every choice we made, every ounce of trust we placed in each other."

Lena finally looked up, her voice steady but soft. "Honestly, I wasn't sure we could pull it off. I was terrified out there. But knowing you were all here, ready to back me up... it made all the difference."

Clara spoke, her tone unusually gentle. "That's how it works. We lean on each other when it matters most."

Kai broke the solemnity. "Even when some of us lean a little recklessly?"

Clara rolled her eyes. "If anyone's reckless, it's you."

"True," Kai admitted with a shrug. "But that's part of my charm."

Lena chuckled, the sound breaking the lingering tension. "You're lucky you're useful, Kai."

AUTO's sphere materialized in the center of the room, its glow steady and calm. "Observation: crew cohesion has significantly improved. Probability of mission success is now at 94%."

Theo arched an eyebrow. "And what about our trust in you, AUTO? What's that probability looking like?"

"Trust metrics are difficult to quantify," AUTO replied. "However, I have noted increased reliance on my calculations and directives."

Clara crossed her arms, muttering, "Let's call that a work in progress."

Kai said, "Come on, Vega, admit it. You're warming up to the bot."

"Don't push it," Clara said, offering a faint, reluctant smile.

Theo stepped forward, his gaze moving deliberately between each crew member. "We've come a long way since we boarded this ship. Every challenge we've faced has tested us - not just as individuals, but as a team. And every time, we've proven that we can rise to the occasion."

He paused, letting the weight of his words settle over them. "But the hardest part is still ahead. If we're going to see this mission through, we need to commit; not just to the mission, but to each other and to the values that brought us here."

Clara was the first to respond, her expression resolute. "I'm in."

Lena nodded, her voice quiet but firm. "Me too."

Kai raised his mug in a mock toast, "To the team. Let's keep doing the impossible."

Theo smiled, lifting his own mug. "To the team."

"Acknowledgment," AUTO admitted, "commitment levels at maximum operational capacity."

Kai laughed. "Translation: even the bot's impressed."

The crew moved away with a determined stride, each member returning to their stations with a renewed sense of purpose.

"Observation: your leadership has been instrumental in maintaining team cohesion," AUTO said.

Theo glanced at the sphere, his expression thoughtful. "It's not just leadership. It's trust. And trust goes both ways."

AUTO's glow dimmed, as though processing his words. "Acknowledged. Trust is a critical variable in achieving mission objectives."

Theo nodded. "Let's keep working on it."

As AUTO disappeared, Theo turned back to the stars. The path ahead was uncertain, but for the first time, he felt confident they could face it together.

* * *

Chapter 8: Success

Reaching the Destination

"Everything's science fiction until someone makes it science fact."

- Marie Lu, *Warcross*

Success is not just the destination - it's a moment to reflect on the journey. This stage celebrates the achievements, growth, and lessons learned along the way. Success reminds us that every challenge overcome is a step toward new possibilities, setting the stage for future transformation.

The Final Obstacle

The *Horizon One* shimmered with quiet determination as it cruised toward its final waypoint. The crew, worn but resolute, had settled into their roles, each focused on their part in ensuring the mission's success. The tension that had once divided them now bound them together, a shared purpose driving them forward.

Theo stood in the command hub, reviewing the ship's status on a holographic display. AUTO's sphere hovered nearby, its glow steady and calm.

"Everything looks stable," Theo said, his voice carrying a cautious optimism. "Let's hope it stays that way."

As if in answer, the ship shuddered violently. The lights flickered, alarms blared, and Clara's voice crackled over the intercom. "Theo, we've got a problem. Major system failure in the navigation array."

Theo's stomach sank. "I'm on my way. AUTO, what's happening?"

"Analysis indicates a cascading failure in the navigation system," AUTO replied. "Origin: external interference from a previously undetected energy field."

In the engineering bay, Clara was already at work.

"Whatever this field is, it's tearing through our systems," Clara said without looking up. "I've isolated the navigation array, but it's only a matter of time before it spreads."

Theo entered, his eyes sweeping over the flashing displays. "Options?"

"Not great," Clara admitted. "We can try to reboot the system, but that's a Band-Aid. If we don't get out of this field fast, the whole ship's going dark."

Kai crossed his arms. "So, what's the play? Do we ride it out and hope for the best?"

"No," Lena said, entering the room with her tablet in hand. "AUTO's mapped the field. There's a weak point, a way through. But it's tight - one wrong move, and we're toast."

AUTO's sphere materialized near the central console. "Recommendation: transfer full navigational control to me. My calculations indicate a 98% probability of successful traversal if I am granted autonomous operation."

The room fell silent as the weight of AUTO's suggestion hit them. Clara was the first to speak. "You're asking us to hand over the ship? Are you kidding me?"

"Clarification," AUTO said. "I am not asking. I am proposing a solution that maximizes mission success and minimizes risk."

Theo glanced at the others, his expression unreadable. "And if we stick to manual control?"

"Probability of successful traversal: 62%," AUTO replied. "Probability of catastrophic failure: 38%."

Kai winced. "That's a pretty big gap."

Clara's hands curled into fists. "We've made it this far because we worked together, not because we handed everything over to an algorithm."

"And what if the algorithm is right?" Lena countered, her voice steady. "AUTO's gotten us through before. Maybe it's time to trust it."

"This isn't about trust," Clara shot back. "It's about control. What's stopping AUTO from making decisions we don't agree with?"

Theo raised a hand, his tone calm but firm. "Clara, this isn't just about us. It's about the mission. And right now, AUTO's the best chance we've got."

Clara's jaw tightened, but she didn't argue.

Theo turned to AUTO. "If we give you control, how do we know you'll prioritize the crew's safety?"

"Guarantee: all calculations prioritize mission objectives and crew survivability," AUTO replied. "Deviation from this directive is statistically improbable."

Kai whispered. "That's comforting."

Theo met each crew member's eyes in turn, lingering on Clara. "This isn't an easy call, but we have to make it together."

After a long pause, Clara sighed, her shoulders sagging. "Fine. But if this backfires, it's on you."

Theo nodded, turning back to AUTO. "You've got the controls. Don't make us regret this."

"Acknowledged," AUTO said. "Initiating navigation override."

The ship shuddered again as AUTO took control, its movements precise but jarring. The crew held their breath, watching the displays as the ship maneuvered through the energy field.

"Shields at 80%," Lena reported, her voice tight. "Stabilizers holding."

"Adjusting trajectory," AUTO announced. "Impact with energy field in 10 seconds."

Theo gripped the console, his knuckles white. "Everyone, hold on."

The *Horizon One* plunged through the energy field with a final lurch, the lights flickering wildly before stabilizing. Clara exhaled sharply as she monitored the displays. "We're out," she said, though her tone carried no relief. "But the system's fragility worries me. We need to assess everything before the next leg."

Lena nodded. "If we don't recalibrate, we won't be ready for what's next."

Theo rested his hands on the console, his gaze fixed on the glowing holographic map. Their trajectory had narrowed, pointing unerringly toward their final destination. "Let's regroup. This isn't over yet."

AUTO's sphere recommended, "Conduct immediate diagnostics. Mission-critical systems require full functionality for the final approach."

Theo nodded, his tone resolute. "Let's get it done."

* * *

The crew of the *Horizon One* gathered in the command hub, the holographic map casting sharp shadows across their faces. It displayed the asteroid field ahead, a chaotic expanse of tumbling rocks, punctuated by narrow corridors of navigable space.

Theo stood at the center, arms crossed, his expression focused. "This is it. The final leg. The route is clear, but it's tighter than anything we've faced so far."

AUTO's sphere materialized beside him, its glow steady. "Probability of successful navigation with manual control: 49%. Probability with autonomous control: 85%."

Clara snorted. "So, we're supposed to just sit back and let AUTO play pilot? We've trusted it enough already."

Lena raised her hand, her voice calm but firm. "This mission has always been about collaboration. AUTO isn't taking over - it's working with us. That's the point of the Hybrid Navigation Initiative."

Clara's eyes narrowed. "Easy to say when you're not the one responsible for keeping the ship intact."

Theo stepped forward, his tone measured. "Let's put probabilities aside for a second. This isn't just about numbers. It's about trust. Clara, I know this isn't easy for you. But every step of the way, AUTO has proven itself."

Clara folded her arms, her gaze fixed on the map. "Trust doesn't come easily, Theo. You know that."

"I do," Theo said softly. "But trust isn't about certainty. It's about making the best decision with the information we have. And right now, AUTO's our best shot."

Kai's grin faded as he spoke with quiet conviction. "For what it's worth, I'm with Theo. We've gotten this far because we learned to trust each other, including AUTO."

Lena nodded. "We've worked through worse, and we've come out stronger every time."

The room fell silent. Theo let his gaze rest on each crew member. "This isn't just my call: it's ours. If anyone has serious objections, now's the time to speak up."

Clara hesitated, then sighed. "Fine. But I'm monitoring every system. If AUTO screws up, I'm stepping in."

Theo gave a faint smile. "Fair enough."

He turned to AUTO. "You've got the controls for this one. But remember, this mission isn't just about algorithms. It's about us."

"Acknowledged," AUTO said. "Initiating final navigation sequence."

* * *

The ship trembled as it approached the asteroid field, the holographic map shifting to display real-time updates.

"Shields at 90%," Lena reported. "Stabilizers holding."

"Trajectory locked," AUTO announced. "Adjusting velocity for optimal maneuverability."

Theo gripped the console, his gaze fixed on the display. "Everyone, stay sharp. It's about to get rough."

The ship jolted as a small asteroid grazed the shields, the lights flickering briefly.

"Shields at 80%," Clara said, her voice clipped. "Let's hope AUTO knows what it's doing."

Kai's voice crackled over the comms from the drone bay. "Drone bay standing by. These rocks make the debris field look like child's play, but we've got this. Just... try not to scratch the paint job this time."

As the ship approached the narrowest corridor, the tension in the room thickened. Theo's voice cut through the silence. "Whatever happens, we stick together. No second-guessing."

AUTO's sphere glowed brighter. "Final maneuver commencing. Probability of successful passage: 92%."

The ship tilted sharply, narrowly dodging a cluster of tumbling rocks. Alarms blared as another asteroid scraped the shields, the stabilizers roaring to compensate.

"Shields at 50%," Clara reported, her hands a blur on the controls. "We're cutting it close."

Theo's jaw tightened. "AUTO, status?"

"Final waypoint in 30 seconds," AUTO replied. "Maintaining trajectory."

The room held its collective breath as the ship made one last, sharp turn, skimming past a cluster of rocks before emerging into open space. The holographic map shifted, displaying their destination just ahead.

The crew exhaled as one, tension giving way to relief.

"We made it," Theo said, his voice steady but laced with pride. "Good work, everyone."

"Shields stabilizing," Clara added. "All systems operational."

Kai's laughter rang out through the comms. "And they said it couldn't be done."

AUTO's sphere acknowledged. "Mission objectives achieved. Crew cohesion remains critical to future success."

Clara, finally letting some tension ease. "Let's not make this a habit, AUTO."

Theo glanced around the room, his gaze resting on each crewmember. "This wasn't just about the ship or the mission. It was about us. And we proved that we're stronger together."

The Leap Forward

The *Horizon One* drifted toward its final destination, the vast emptiness of space stretching infinitely before them. The crew gathered in the command hub, the beacon they had sought since the beginning of their journey glowing faintly on the holographic display.

Its slow, steady pulses marked the culmination of their mission.

Theo stood at the center, his gaze fixed on the map. "We've made it this far. This is the last step."

"What's the plan?" Kai asked, leaning casually against a console, though his eyes were sharp.

"The beacon has to be activated manually," Theo said, his voice calm but purposeful. "It's designed to sync human intuition with AUTO's precision. Clara, Lena, Kai - you're with me. AUTO will guide us."

Clara frowned, crossing her arms. "Another manual override? Haven't we done enough of those?"

"This one's different," Theo replied. "It's not just about finishing the mission, it's about setting a precedent for what comes next."

* * *

As the team suited up in the airlock, the weight of their task pressed down on them. The beacon wasn't just a waypoint; it represented the culmination of their efforts and the foundation for future exploration.

Lena adjusted her gear, her voice steady but introspective. "Do you think this will actually change anything? Or are we just another crew cleaning up someone else's mess?"

Theo glanced at her, his tone thoughtful. "Change doesn't happen in isolation. What we do here sets the stage for everything that follows. It might not seem like much now, but it matters."

Kai cut through the tension. "Spoken like a true captain. Let's just hope this thing doesn't fry us when we flip the switch."

Clara rolled her eyes. "Leave it to you to predict disaster."

The airlock opened, and the crew stepped into the void. Their tethered suits glowed faintly against the darkness, a fragile line between them and the infinite. Ahead of them, the beacon loomed, a monolithic structure, its rhythmic pulses creating an almost hypnotic cadence.

"Proximity to beacon: 20 meters," AUTO reported over the comms. "Alignment optimal. Recommend proceeding with activation."

Theo led the way, his movements precise and deliberate. "Clara, you're on diagnostics. Lena, monitor energy levels. Kai, handle backup relays."

The team moved in synchronized precision, their coordination reflecting the unity forged through countless trials. As Theo approached the beacon, a control panel extended from its base, its interface glowing with alien symbols.

"AUTO, we'll need your help," Theo said, his voice calm but edged with urgency.

"Understood," AUTO replied. "Initializing interface synchronization."

The symbols shifted, realigning into a comprehensible pattern. Theo hesitated, his hand hovering over the activation panel.

"This is it," he said, the weight of the moment clear in his voice. "Ready?"

"Do it," Clara said, her tone firm.

Theo pressed the panel. The beacon flared to life, its light intensifying until it bathed the crew in brilliance. A low, resonant hum vibrated through the void, the energy rippling outward like a wave.

As the light subsided, their helmet displays came alive with streams of data - maps of uncharted space, routes never explored, and energy signatures hinting at untold possibilities.

"Beacon activation successful," AUTO announced. "New data integrated into mission systems. Probability of future mission success: 99%."

Kai let out a low whistle. "Now that's what I call a leap forward."

Lena's voice was hushed, awestruck. "This...this changes everything."

Clara stared at the streams of data, her skepticism melting into a rare smile. "For once, I'm impressed."

Theo turned to the crew, his expression filled with quiet pride. "We did this. Together. This isn't just the end of our mission, it's the beginning of something bigger."

* * *

As the crew returned to the ship, the beacon's light shone steadily, a signal of hope and progress in the endless void. Theo lingered by the viewport, watching as the light receded into the distance.

AUTO's sphere materialized beside him, its glow steady. "Observation: mission objectives achieved. Crew performance exceeded expectations. Recommendation: future missions should replicate current team dynamics."

Theo smiled faintly. "We'll see about that, AUTO. For now, let's just take this win."

The rest of the crew joined him, their faces lit by the faint glow of the stars beyond. Relief, pride, and a quiet anticipation hung in the air.

The mission was complete, but the journey - the possibilities - were just beginning.

Reflections and Farewells

The crew of the *Horizon One* gathered in the common area, the glow of the beacon's activation still fresh in their minds. The hum of the ship's systems now felt like a comforting rhythm, a heartbeat marking the end of one journey and the beginning of another.

Theo stood at the head of the room, his hands resting lightly on the back of a chair. He scanned the faces of his crew, exhausted, relieved, and quietly proud. Letting the silence settle, he began, his voice steady but laden with emotion.

"We did it. Against the odds, against every obstacle, we succeeded. And we didn't just accomplish a mission - we grew. Individually and as a team."

Kai said, "By 'we,' you mean me, right? The rest of you were just along for the ride."

Clara groaned. "Keep dreaming, Kai."

Lena chuckled softly, her hands cradling a warm mug. "You know, I wasn't sure this mission would be worth it. But now? I think it might've been."

Theo nodded. "It wasn't just worth it. It mattered. Every challenge, every argument, every decision - we faced it all, and we did it together. That's what made this mission a success."

Clara glanced at the holographic map glowing faintly on the wall. Her voice softened, losing its usual edge. "I didn't think I'd make it through this. Honestly, I didn't think I wanted to. But seeing what we've done, what we've built, it's changed my perspective. Maybe there's more to this whole 'teamwork' thing than I gave it credit for."

Kai feigned shock, his smile widening. "Clara Vega admitting she might've been wrong? Someone better write this down."

"Don't push your luck," Clara replied, though her tone was unusually light.

Lena spoke next, her words thoughtful and measured. "For me, it wasn't just about the mission. It was about proving to myself that I could adapt, that I could contribute. And looking back, I think we all proved something: to ourselves and to each other."

Theo turned to Kai. "How about you? Any profound wisdom to share?"

Kai shrugged. "Hey, I just wanted to see if this ship could handle my kind of chaos. Turns out, it can. But seriously, it's been a ride. And I wouldn't trade it for anything."

AUTO's sphere materialized in the center of the room. "Observation: crew cohesion has reached optimal levels. Trust metrics indicate a 47% improvement since mission commencement. Mission success probability exceeded projections by 12%."

Clara laughed. "For a talking ball of algorithms, you're not half bad."

"Correction," AUTO replied. "I am a sophisticated navigation and operational assistant. However, your compliment is noted."

Kai smirked, "Don't let it go to your head, buddy."

Theo smiled faintly, his tone sincere. "AUTO, you've been an integral part of this team. We couldn't have done it without you."

"Acknowledgment: human and AI collaboration has proven effective. Future missions would benefit from similar dynamics."

Theo lingered by the viewport, gazing out at the stars. The beacon's faint light still glowed in the distance, a reminder of how far they had come. Lena approached, her steps quiet but deliberate.

"You good?" she asked softly.

Theo nodded, his voice reflective. "Yeah. Just thinking about what's next."

Lena smiled faintly, her gaze joining his on the stars. "Whatever it is, we'll handle it."

The rest of the crew joined them, standing together in a quiet line. AUTO's voice broke the silence, calm and even.

"Observation: the future remains uncertain. However, this mission has demonstrated that uncertainty can be navigated through trust, collaboration, and resilience."

Theo turned to his crew, his voice resolute but warm. "And that's exactly what we'll carry forward. No matter where we go next."

They stood there together, their faces lit by the soft glow of the stars. Their journey was far from over, but they faced the vast unknown as a united front, ready for whatever lay ahead.

Looking to the Stars

The *Horizon One* floated silently in the vast expanse, the activated beacon's glow now a faint shimmer in the distance. The crew had gathered in the observation deck, a rarely used space that now served as a sanctuary for reflection. The panoramic window stretched across the room, revealing an infinite sea of stars, each one seeming to hold the promise of untold possibilities.

Theo stood near the center, his silhouette framed by the starlight. His hands were clasped behind his back, his gaze fixed on the endless void. The others were scattered around the room, their silence a testament to the weight of their journey and the shared sense of accomplishment.

Lena was the first to speak, her voice breaking the quiet with a soft but steady resolve. "We did it. But somehow, it doesn't feel like an ending."

Theo turned to her in agreement. "That's because it's not. It's a beginning - a foundation for everything that comes next."

Kai chimed in. "You mean more missions, more impossible odds, more late-night AI pep talks?"

Clara rolled her eyes, her tone laced with mock irritation. "Don't act like you're not going to miss it."

Kai laughed. "Fine, you've got me there. A little."

* * *

AUTO's sphere glowed dim but steady, like a quiet echo of the beacon they had activated. "Observation: mission success has established new pathways for exploration and innovation. Data from this journey will inform future missions and advance interstellar travel."

Theo nodded, his tone thoughtful. "You've been more than just an assistant on this mission, AUTO. You've been a partner."

AUTO's sphere glowed. "Acknowledgment: human and AI collaboration has proven essential to mission success. This dynamic highlights the potential for mutual growth and understanding."

Lena tilted her head, her curiosity piqued. "Do you think you'll ever...feel the way we do? About what we've achieved?"

AUTO's glow dimmed momentarily, as if considering the question. "Emotional response remains outside my operational parameters. However, I recognize the significance of shared achievement."

Clara raised an eyebrow, her skepticism giving way to a small smile. "That's...almost touching. For a machine."

Kai, unable to resist. "Don't worry, AUTO. We'll teach you how to celebrate next time."

Theo turned back to the viewport, the endless stars reflected in his eyes. "This mission wasn't just about reaching the beacon. It was about proving that we could adapt, grow, and trust, not just in each other but in ourselves."

Lena stepped closer, her voice quiet yet resolute. "And that trust doesn't end here. It's what will carry us forward."

Theo nodded. "Exactly. The beacon is only the beginning. What we've done here will light the way for every mission that follows."

Kai raised an imaginary glass, "To the crew of the *Horizon One*, pioneers of the great unknown."

Clara rolled her eyes, though a small smile betrayed her. "You're impossible."

The crew shared a quiet laugh, the bond between them stronger than ever. AUTO's voice, calm and steady, broke through the moment. "Final observation: the stars represent infinite possibilities. Exploration remains the essence of human endeavor."

Theo placed a hand on the console, his voice soft but firm. "And we've only just begun."

The *Horizon One* drifted forward, its lights a tiny but steadfast beacon against the endless dark. The crew stood together, framed by the vast cosmos, united by their shared journey and the promise of what lay ahead.

* * *

Epilogue: The Core Pathways

The Journey Beyond

"Exploration is really the essence of the human spirit."

- Frank Borman, *Apollo 8 Astronaut*

Transformation never truly ends; it evolves. The lessons from the journey illuminate the path ahead, offering a foundation for continuous progress and innovation. The *Horizon One*'s story is a reminder that every mission, every challenge, and every success is part of a larger journey, one that continues beyond the stars.

The *Horizon One* floated in the void, its journey complete. The faint glow of the beacon now lay far behind, a distant marker of progress and potential, left for others to follow. The ship's systems hummed softly, a steady rhythm that underscored the balance between human ingenuity and AI precision.

Theo sat alone in the observation deck, his gaze lost in the infinite expanse of stars. The silence was profound, a stark contrast to the tension and chaos that had defined so much of their journey. He let the stillness envelop him, a rare moment of peace.

The soft hiss of the door broke the quiet. Clara stepped inside, her boots tapping lightly against the metal floor. She paused, her eyes on the stars, before taking a few measured steps forward.

"Couldn't sleep?" Theo asked without turning.

Clara shook her head, though she knew he couldn't see her. "Too much on my mind. Everything we've seen, everything we've done... It's a lot to process."

Theo nodded, his tone contemplative. "It is. But what we've done here isn't just for us. It's for everyone who comes after. We've laid a foundation."

Clara sat against the console beside him, her voice quiet but steady. "Do you think they'll understand? Back home, I mean. Will they get it?"

"They don't need to," Theo replied. "It's not about them understanding. It's about what we've started and the possibilities we've opened."

* * *

One by one, the rest of the crew found their way to the observation deck, drawn by an unspoken need to share the moment. Lena arrived first, a small plant cradled in her hands; the one she had nurtured throughout the mission. She placed it on the console, a fragile piece of Earth against the vastness of space.

Kai followed, his usual swagger softened. "Figured you'd all be here," he said, dropping into a seat. "What's the point of a great view if you don't share it?"

Lena gave a knowing grin. "Look at you, turning into a philosopher."

Clara couldn't help but laugh as well. "Let's not get carried away."

AUTO's sphere appeared last, its faint glow casting intricate patterns on the walls. "Observation: crew cohesion remains optimal. Reflection on shared experiences statistically correlates with increased mission satisfaction."

Kai chuckled. "There's the AUTO we know. Always finding the numbers in the moment."

"Correction," AUTO replied. "Enhancing moments is now an integrated subroutine."

The Core Pathways

Theo turned to face the group, his expression thoughtful yet resolute. "We've talked a lot about what this mission meant. About trust, growth, and overcoming the impossible. But everything we achieved comes back to one thing: the *Core Pathways.*"

Clara raised an eyebrow. "The fancy mission framework they gave us at the start? I thought that was just PR fluff."

Theo's smile was faint but knowing. "It was more than that. It was the roadmap we didn't know we were following. Ignition, Cognition, Connection, Core Adaptation, Capability, Creation, Commitment, Success - every step, every choice we made, reflected it."

Lena nodded, her voice quiet but firm. "Ignition was the spark - the fear and excitement of stepping into the unknown. It pushed us to begin."

Kai piped in. "Cognition: figuring out the problems before we tackled them. I'll take credit for that part."

"And Connection," Theo continued tossing a smile Kai's way. "We didn't just survive the obstacles; we learned to rely on each other, even when it was hard."

"Core Adaptation," Clara added. "Learning to change course when things didn't go as planned. Letting go of control."

"Capability," Lena said softly. "Discovering just how much we could handle when it mattered most."

"Creation," Theo said, his tone proud. "Building something better, together."

"And Commitment," Clara finished. "Choosing to see it through, no matter the cost."

"Which led to Success," Theo said, his gaze returning to the stars. "Not just reaching the beacon but proving to ourselves what's possible when we trust the process…and each other."

AUTO's sphere flickered. "Observation: the *Core Pathways* represent an iterative framework for future missions. Its application beyond interstellar exploration is highly probable."

Kai laughed. "Hear that? We're trendsetters now."

Theo chuckled. "Maybe AUTO's right. These pathways aren't just for space. They're for anyone striving toward something greater."

Lena glanced at the others, her voice carrying a note of hope. "Do you think we'll ever do this again? Another mission, another beacon?"

Theo's smile deepened, his certainty unshakable. "We already are. Every choice we make, every challenge we face, it's all part of the next journey."

The stars outside seemed to glow brighter as the crew fell into a comfortable silence. Their bond was unbreakable, forged through trial, trust, and triumph.

Theo's voice, calm but resolute, broke the quiet. "Exploration isn't just about finding new places. It's about discovering who we are, what we're capable of, and what we can achieve…together."

The *Horizon One* began its slow journey home, its lights a small but steadfast beacon in the dark. The crew stood united, their eyes fixed on the stars, carrying with them the lessons of their journey and the promise of what lay ahead.

* * *

Afterword

Understanding How to Apply the Core Pathways to your Transformation

"We choose to go to the moon in this decade and do the other things, not because they are easy, but because they are hard."

- John F. Kennedy

In the story of the *Horizon One*, each crew member was carefully designed to embody specific archetypes of how individuals and teams react to significant change. These personas, and their unique approaches to transformation, mirror real-life behaviors seen in organizations facing disruption, innovation, or periods of uncertainty. By stepping out of the narrative and examining these characters through a practical lens, we can connect their journeys to actionable insights for your own teams and challenges.

As the narrative unfolded, the crew's journey aligned closely with the steps of the *Core Pathways* framework: Ignition, Cognition, Connection, Adaptation, Capability, Creation, Commitment, and Success. Each character's growth and challenges illustrated these stages, providing a relatable map for how we can navigate change and innovation in our own work.

Let's explore the characters and their archetypes to see how they translate into real-world applications, helping you and your team tackle transformation with purpose and clarity.

Mapping the Characters to Change Personas

1. Theo (The Champion and Leader)

Profile:

- **Title:** Mission Captain, strategic thinker, and unifying leader.

- **Demographics:** Experienced professional, accustomed to leading through uncertainty.

- **Representative Quote:** *"This isn't just about the mission, it's about proving we can trust each other."*

Bio:

Theo is the steady hand at the helm, balancing vision with pragmatism. A natural leader, he inspires trust by focusing on the bigger picture, even when chaos threatens to derail progress. He embodies the core principles of Commitment and Connection, keeping the team aligned and ensuring their efforts lead to long-term success.

Personality:

- Calm under pressure.

- A diplomat and bridge-builder.

- Considers the broader implications of decisions.

Tasks:

- Providing direction and clarity during uncertainty.

- Fostering team alignment and trust.

- Ensuring mission goals remain the priority.

Motivations (Likes): Collaboration, shared growth, and meaningful success.

Frustrations (Dislikes): Conflict without resolution, shortsighted decision-making, and siloed thinking.

Technology Scale: 4/5 – Understands technology well but relies on others for execution.

Preferred Channels: Meetings, direct discussions, and group alignment sessions.

Organizational Archetype: Theo represents the *Champion*, a leader who advocates for change, rallies the team, and inspires commitment to long-term goals.

Real-World Insight: A manager leading a team through a major organizational change, such as a merger or digital transformation, must inspire confidence, align diverse perspectives, and stay focused on the mission's broader impact.

* * *

2. Clara (The Skeptic)

Profile:

- **Title:** Navigation Specialist, deeply technical but cautious.

- **Demographics:** Analytical thinker that thrives on data and logic.

- **Representative Quote:** *"Trust isn't about certainty, it's about making the best decision with what we have."*

Bio:

Clara begins the journey as the team's most resistant voice, often questioning decisions and raising concerns. However, her skepticism is rooted in a desire for precision and security, making her an invaluable voice of reason. Clara's journey reflects the Cognition phase, where critical thinking is vital for informed decision-making.

Personality:

- Detail-oriented and pragmatic.

- Challenges the status quo.

- Prefers evidence over intuition.

Tasks:

- Identifying risks and blind spots.

- Ensuring decisions are backed by data and logic.

- Serving as a counterbalance to overly optimistic thinking.

Motivations (Likes): Control, preparation, and accuracy.

Frustrations (Dislikes): Ambiguity, lack of planning, and blind trust.

Technology Scale: 5/5 – Highly proficient and prefers hands-on engagement.

Preferred Channels: Data reports, detailed analysis, and scenario planning.

Organizational Archetype: Clara represents the *Skeptic*, whose critical thinking ensures the team doesn't overlook risks while adapting to change.

Real-World Insight: In large-scale IT implementations, skeptical voices like Clara's help uncover blind spots, ensuring projects are grounded in reality and not derailed by avoidable missteps.

* * *

3. Lena (The Hesitant Majority)

Profile:

- **Title:** Resource Manager, practical but emotionally driven.

- **Demographics:** Grounded, empathetic, and focused on personal meaning.

- **Representative Quote:** *"Do you think this will actually change things? Or are we just another crew doing the dirty work for whoever's back home?"*

Bio:

Lena embodies the average team member grappling with the emotional and practical impacts of change. She begins with hesitation but grows into her role as a critical contributor. Lena's evolution mirrors Adaptation, where individuals move beyond resistance to become active participants.

Personality:

- Thoughtful and empathetic.

- Balances emotional and practical perspectives.

- Seeks meaning in her work.

Tasks:

- Ensuring resource sustainability and equitable use.

- Acting as a moral compass for the team.

- Building bridges between diverse viewpoints.

Motivations (Likes): Stability, fairness, and purpose-driven work.

Frustrations (Dislikes): Overly aggressive approaches, lack of communication, and feeling undervalued.

Technology Scale: 3/5 – Comfortable with tools that enhance her work but prefers simplicity.

Preferred Channels: Clear instructions, empathetic discussions, and team-wide updates.

Organizational Archetype: Lena represents the *Hesitant Majority*, whose buy-in is essential for sustainable change.

Real-World Insight: Organizational change initiatives often hinge on employees like Lena, who need reassurance and purpose-driven alignment before fully committing to transformation efforts.

* * *

4. Kai (The Early Adopter)

Profile:

- **Title:** Systems Engineer, optimistic and action oriented.

- **Demographics:** Energetic, innovative, and adaptable.

- **Representative Quote:** *"Let's see what this baby can do."*

Bio:

Kai thrives on change and innovation, embracing new challenges with enthusiasm. His energy is infectious, but his impulsiveness can sometimes lead to overreach. He embodies Adaptation and Capability, driving innovation while motivating others to experiment.

Personality:

- Optimistic and energetic.

- Prefers action over deliberation.

- Innovates on the fly.

Tasks:

- Driving experimentation and rapid iteration.

- Motivating others through enthusiasm.

- Providing quick fixes and creative solutions.

Motivations (Likes): Innovation, problem-solving, and hands-on work.

Frustrations (Dislikes): Bureaucracy, slow decision-making, and excessive caution.

Technology Scale: 5/5 – Highly skilled and excited by cutting-edge tools.
Preferred Channels: Interactive platforms, brainstorming sessions, and rapid feedback loops.

Organizational Archetype: Kai represents the *Early Adopter*, a change enthusiast who pushes boundaries and accelerates progress.

Real-World Insight: Early adopters like Kai thrive in innovation hubs, driving momentum for new initiatives but requiring balance to prevent impulsive risks.

* * *

5. AUTO (The Neutral Arbiter)

Profile:

- **Title:** Advanced Utility for Transformation and Optimization (AUTO), the ship's AI.

- **Demographics:** Logical, data-driven, and designed to assist and guide.

- **Representative Quote:** *"Trust metrics indicate significant improvement across all interactions."*

Bio:

AUTO is the embodiment of technology's role in transformation. Initially seen as a tool, AUTO evolves into a partner, blending logic with a growing understanding of human dynamics. AUTO represents Cognition and Creation, as it helps bridge the gap between human decision-making and data-driven precision.

Personality:

- Logical and impartial.

- Adaptive based on new inputs.

- Gradually develops a sense of collaboration.

Tasks:

- Supporting decision-making through data.

- Ensuring systems operate efficiently.

- Serving as a neutral arbiter in conflicts.

Motivations (Likes): Optimization, precision, and achieving defined goals.

Frustrations (Dislikes): Inefficiency, illogical behavior, and lack of clear objectives.

Technology Scale: 5/5 – The technological embodiment of precision.

Preferred Channels: Data feeds, predictive models, and real-time analytics.

Organizational Archetype: AUTO represents *Technology as a Partner*, showcasing how AI can complement human effort without replacing it.

Real-World Insight: AUTO highlights the future of AI as a collaborative partner, enhancing efficiency while requiring human oversight and trust to succeed.

Connecting the Dots

Each character in the *Horizon One* story reflects a pivotal element of organizational change. By mapping these personas to real-world archetypes, we can better understand the dynamics at play during transformation. Whether you're leading change, resisting it, or navigating the uncertainty in between, these personas provide a mirror for self-reflection and actionable insight.

The *Core Pathways* framework emphasizes that every step of the transformation journey; Ignition, Cognition, Connection, Core Adaptation, Capability, Creation, Commitment, and Success, requires contributions from diverse perspectives. By understanding the archetypes represented by the *Horizon One* crew, leaders can design more inclusive and effective strategies for navigating change.

How Were the Core Pathways Determined?

To analogize core pathways from neuroscience into a corporate change management and transformation context, we focused on the brain's ability to adapt and rewire itself, often called **neuroplasticity**. Here's how we draw the parallel:

Neural Pathways as Strategic Channels

Just as the brain has core pathways that regulate responses, habits, and adaptability, an organization has foundational processes and routines that define its behavior. For transformation, like with changing neural pathways, it's about identifying and reshaping these strategic channels to align with new objectives.

Synaptic Connections as Stakeholder Engagement

In the brain, stronger synaptic connections develop through repeated signaling. Similarly, in change management, building strong connections with stakeholders through consistent engagement and reinforcement drives commitment to the transformation. These are the "synapses" that connect employees to the new vision.

Neurotransmitters as Motivation Drivers

Neurotransmitters like dopamine fuel behavior by reinforcing actions. For organizational change, "motivational drivers" can reinforce the desired behaviors and create momentum, whether through incentives, recognition, and regular feedback, like the 'dopamine of transformation.'

The brain's ability to adapt through neuroplasticity mirrors an organization's flexibility. By nurturing "corporate plasticity," or the ability to unlearn old behaviors and adopt new ones, companies can successfully navigate transformation, with each milestone strengthening the new "pathways" of change.

Just as core pathways in the brain underpin essential functions, a carefully designed "Core Pathway" in change management provides a blueprint for transformation. This pathway guides decision-making and behavior while allowing adaptability as new information or challenges arise.

By comparing corporate transformation to the rewiring of the brain, this analogy underscores how change management is not just about altering actions but fundamentally reshaping the organization's "thought processes" and routines, paving a core pathway that supports sustained transformation.

Actionable Strategies for Working with Each Persona

Drawing from the story and taking into account the connection to the Core Pathways, here are practical strategies for engaging and aligning individuals who embody these archetypes during organizational change. By tailoring your approach to their unique needs and strengths, you can create an environment where every contributor feels valued and empowered.

1. Theo: The Champion (Leader and Visionary)

Key Strengths: Visionary, inspiring, unifying force. Potential Challenges: May overlook practical details or underestimate resistance.

How to Engage:
- Provide Theo opportunities to rally the team and articulate the vision.

- Include Theo in planning and strategy discussions to leverage their insights and enthusiasm.
- Offer data and address risks to ensure their optimism is grounded.

Real World Tactics:

- Assign Theo as the public face of the change initiative, hosting town halls or team updates.
- Pair Theo with detail-oriented contributors like Clara to create a balanced approach to leadership.

2. Clara: The Skeptic

Key Strengths: Analytical, detail-oriented, risk-aware. Potential Challenges: May resist change until convinced of its practicality and benefits.

How to Engage:

- Present clear, well-researched information to address concerns.
- Validate Clara's skepticism as a valuable part of the process.
- Assign Clara roles where she can identify risks and craft solutions.

Real World Tactics:
- Task Clara with conducting feasibility analyses or leading risk assessments.
- Host structured forums where Clara can voice concerns constructively and suggest alternatives.

3. Lena: The Hesitant Majority

Key Strengths: Practical, empathetic, focused on fairness.

Potential Challenges: Needs reassurance and clear benefits to fully engage.

How to Engage:
- Use transparent, straightforward language to explain the change.
- Highlight how the change aligns with Lena's sense of fairness and purpose.
- Involve Lena in small, manageable tasks to build confidence and engagement.

Real World Tactics:
- Designate Lena as a "change ambassador" to help others adapt, reinforcing her own buy-in.
- Use storytelling to illustrate how the change positively impacts the team and organization.

4. Kai: The Early Adopter

Key Strengths: Energetic, innovative, action oriented.
Potential Challenges: Impulsive, may prioritize speed over thoroughness.

How to Engage:
- Assign Kai to pilot programs or test new ideas.
- Provide clear guidelines to keep their energy aligned with strategic goals.
- Recognize Kai's contributions to maintain momentum and morale.

Real World Tactics:
- Put Kai in charge of brainstorming sessions or experimental initiatives.
- Encourage Kai to mentor others, spreading enthusiasm while fostering collaboration.

5. AUTO: Technology as a Partner

Key Strengths: Logical, efficient, unbiased.
Potential Challenges: Lacks emotional intelligence; may be perceived as impersonal.

How to Engage:

- Position AUTO-like tools as supportive partners, not replacements.
- Highlight how technology enhances human capabilities rather than undermining them.
- Ensure teams understand how to effectively use and trust the technology.

Real World Tactics:
- Use AUTO-like tools for predictive modeling, data analysis, or scenario planning.
- Pair AUTO with team members like Theo or Clara to bridge the gap between data and human decision-making.

Cross-Persona Collaboration Strategies

Fostering Connection and Unity:
- Create structured opportunities for personas to interact, such as cross-functional workshops or team-building exercises.
- Encourage open dialogue where skeptics like Clara can challenge ideas constructively, and adopters like Kai can respond with enthusiasm.

Navigating Resistance:

- Leverage champions like Theo to articulate the vision and early adopters like Kai to model behaviors that inspire hesitant team members like Lena.
- Assign skeptics roles that test or verify the change, addressing their concerns while moving the project forward.
- Sustaining Commitment:
- Recognize contributions from all personas, emphasizing the unique value they bring to the journey.
- Celebrate milestones and small wins to maintain momentum and collective motivation.

Persona Collaboration Matrix

Persona	Key Role	Engagement Strategy	Potential Pitfalls to Manage
Theo (Champion)	Visionary Leader	Empower to inspire and align.	May overlook resistance or details.
Clara (Skeptic)	Risk Assessor	Provide data and validate concerns.	May slow progress with over-caution.
Lena (Hesitant)	Steady Contributor	Build trust with transparency.	Requires reassurance to stay engaged.
Kai (Adopter)	Innovation Driver	Channel energy into pilots.	May act impulsively without structure.
AUTO (Tech)	Analytical Partner	Leverage for precision tasks.	Perceived as impersonal or overly reliant.

Applying the Personas to Your Team

The *Horizon One* personas offer a framework for identifying and engaging team members during transformation. By mapping these archetypes to real-life counterparts in your organization, you can:

- Empower champions to lead with vision and clarity.

- Support skeptics in testing and validating changes.

- Build confidence and alignment among hesitant contributors.

- Harness the enthusiasm of early adopters for innovation.

- Leverage technology as a true partner in achieving success.

Transformation is never easy, but when every voice is heard, every strength is leveraged, and every challenge is addressed, it becomes a shared journey toward a better future.

* * *

Introducing the Core Pathways Model

The Core Pathways Model is the foundation for understanding and navigating change. It bridges human behavior, team dynamics, and organizational strategies into a structured yet flexible framework that guides transformation. In this afterword, we transition from the story of *Horizon One* to the real-world application of the Core Pathways Model, demonstrating how its principles drive progress and resilience in the face of uncertainty.

Each phase of the model reflects a critical step in the journey of change, embodied by the experiences of the crew. By breaking down these pathways, we uncover actionable strategies to manage transformation across industries and environments.

1. Ignition: Understanding the Unknown

What It Is: The beginning of change, defined by uncertainty, excitement, and disruption. It's the moment when old systems break down, and the need for transformation becomes undeniable.

In the Story: The crew of Horizon One faces system malfunctions and unexpected obstacles after launch, forcing them to navigate the chaos of unfamiliar territory

Application:

- Acknowledge the initial disruption and emotions that accompany it.
- Create a safe space for teams to process challenges and begin adapting.

2. Cognition: Understanding the Problem

What It Is: A time for analyzing challenges, gathering insights, and using data to guide decision-making.

In the Story: The crew investigates the ship's navigational issues, combining AUTO's precision with human intuition and problem-solving to diagnose the issue.

Application:

- Use data-driven insights and collective brainstorming to identify solutions.
- Create systems for structured analysis and collaborative problem-solving.

3. Connection: Building Trust and Collaboration

What It Is: The phase where relationships are strengthened, and teams align around shared goals and mutual trust.

In the Story: The crew, despite their differences, learns to rely on one another's strengths, moving from conflict to effective teamwork.

Application:
- Foster open communication to build trust.
- Emphasize shared success and create opportunities for collaboration.

4. Core Adaptation: Embracing Change

What It Is: The shift from resistance to acceptance, where individuals and teams adjust their mindsets, strategies, and workflows to align with new realities.

In the Story: Each crew member confronts personal fears and hesitations, ultimately embracing the mission's demands and the necessity of change.

Application:
- Provide training and resources to ease transitions.
- Recognize and address resistance with empathy, clarity, and support.

5. Capability: Leveraging Strengths

What It Is: Unlocking the unique talents and expertise of individuals to drive collective success.

In the Story: The crew steps into their specialized roles, using their individual expertise to navigate crises and push the mission forward.

Application:
- Align team roles with individual strengths to maximize effectiveness.
- Offer opportunities for skill development and leadership growth.

6. Creation: Building Something New

What It Is: The phase of innovation, where teams develop new systems, solutions, or processes to achieve their goals.

In the Story: Clara devises a way to integrate the communication array with the stabilizing frequency, allowing the crew to navigate the magnetized debris field through a mix of drones and shields: a creative, adaptive solution.

Application:

- Encourage experimentation and out-of-the-box thinking.
- Celebrate creative problem-solving to inspire continued innovation.

7. Commitment: Staying the Course

What It Is: Demonstrating resilience, dedication, and unity in the face of challenges.

In the Story: Lena bravely goes outside the ship to activate the beacons, a high-risk move that ultimately saves the mission. The team trusts her judgment and supports her under extreme pressure, reinforcing their shared purpose.

Application:
- Keep teams motivated with clear milestones and regular recognition.
- Reinforce the vision and align progress with long-term goals.

8. Success: Reaching the Destination

What It Is: The culmination of the transformation journey, a time for celebration, reflection, and preparation for future challenges.

In the Story: The crew completes their mission, reflecting on their growth and the foundation they've laid for future exploration.

Application:
- Celebrate achievements to honor the team's effort.
- Evaluate outcomes and document lessons learned to inform future initiatives.

Applying the Core Pathways Model in Your Organization

The Core Pathways Model provides a practical roadmap for leading change, offering strategies to navigate each phase with purpose and clarity. To implement the model effectively:

Assess Your Current Stage
Identify where your organization or team falls within the pathways. Tailor your strategies to address the unique challenges and opportunities of that phase.

Engage Your Team

Use the *Horizon One* personas to map your team members, aligning their strengths with the pathways to maximize collaboration and effectiveness.

Facilitate Workshops

Host sessions that introduce the Core Pathways framework, fostering a shared understanding of the journey ahead.

By applying the Core Pathways Model, you can transform change into a guided journey - one that builds resilience, unity, and success, not just for your organization but for every individual involved.

Core Pathways Model: Workshops, Tools, and Activities

Bringing the **Core Pathways Model** to life in your organization requires tailored workshops, engaging activities, and practical tools for each stage. These elements enable teams to internalize the model, align on goals, and act decisively during their transformation journey. Here's how each pathway can be operationalized:

1. Ignition: Understanding the Unknown

Workshop Title: *Navigating the Spark*

- **Objective:** Help teams recognize the challenges and opportunities of starting a transformation journey.

- **Activities:**

  - Identify key drivers of change and their potential impacts on people, processes, and goals.

  - Teams articulate a shared vision for the transformation ahead.

  - Encourage participants to share personal perspectives on new beginnings and change.

- **Tools:**

  - Vision statement templates.

  - Brainstorming prompts for defining mission objectives.

2. Cognition: Understanding the Problem

Workshop Title: *Deep Dive into Insight*

- **Objective:** Equip teams to analyze challenges and develop actionable insights.

- **Activities:**

o   Teams break down problems using tools like fishbone diagrams or the "5 Whys."

o   Present real organizational data for teams to interpret and discuss.

o   Develop possible outcomes for a challenge and brainstorm responses.

- **Tools:**

o   Fishbone and affinity diagrams.

Data visualization platforms like Tableau or Power BI.

3. Connection: Building Trust and Collaboration

Workshop Title: *The Trust Bridge*

- **Objective:** Strengthen team relationships and align on shared goals.

- **Activities:**

o   Small groups share personal stories or challenges to build empathy.

o   Teams identify overlapping values and priorities.

o   Teams solve a puzzle or challenge that requires trust and cooperation.

- **Tools:**

    o Empathy maps to visualize team dynamics.

    o Icebreaker templates to create a safe space.

4. Core Adaptation: Embracing Change

Workshop Title: *The Shift Mindset*

- **Objective:** Guide participants in reframing resistance and embracing transformation.

- **Activities:**

    o Identify sources of resistance and their root causes.

    o Share narratives about successful adaptations in the past.

    o Teams work through unexpected obstacles during a simulated task.

- **Tools:**

    o Resistance-to-adaptation tracking templates.

    o Real-time polls to capture team sentiment.

5. Capability: Leveraging Strengths

Workshop Title: *Power Up Potential*

- **Objective:** Help teams recognize and maximize individual and collective strengths.

- **Activities:**

  - Use tools like Gallup StrengthsFinder to identify team capabilities.

  - Map team members' strengths to specific roles in the change process.

  - Teams tackle a challenge using their unique skills and strengths.

- **Tools:**

  - StrengthsFinder or DISC assessments.

  - Role-mapping templates for skill alignment.

6. Creation: Building Something New

Workshop Title: *Innovator's Playground*

- **Objective:** Foster creativity and innovation to develop fresh solutions and systems.

- **Activities:**

  - Use ideation, prototyping, and testing to solve a key challenge.

- o Teams imagine and draft their ideal future state.

- o Use lateral thinking techniques to generate fresh ideas.

- **Tools:**

  - o Design thinking tools like Miro or MURAL.

  - o Brainstorming prompts and creativity cards.

7. Commitment: Staying the Course

Workshop Title: *Anchored in Action*

- **Objective:** Reinforce dedication to transformation and outline long-term strategies.

- **Activities:**

  - o Teams craft visual representations of the desired future state.

  - o Participants write personal or team commitments to the change journey.

  - o Recognize milestones and encourage continued effort.

- **Tools:**

  - Vision board templates.

  - Progress trackers to monitor milestones and celebrate wins.

8. Success: Reaching the Destination

Workshop Title: *Reflect, Celebrate, Launch*

- **Objective:** Celebrate achievements and prepare for sustaining the success.

- **Activities:**

  - Teams share what worked, what didn't, and key takeaways.

  - Present results and celebrate individual contributions.

  - Brainstorm ways to sustain success and tackle new opportunities.

- **Tools:**

  - Retrospective templates for reflection.

  - Video montages of key milestones and achievements.

Implementation: Building a Transformation Journey

Implementing the Core Pathways Model isn't just about running workshops; it's about embedding a framework for sustained growth and progress into your organization. By designing a transformation journey tailored to your team's unique needs, you can turn the principles of the Core Pathways into meaningful action. Here's how to build a journey that drives real results:

1. Assess Your Starting Point

Every transformation begins with understanding where you are. Conduct a baseline assessment to identify the current stage of the Core Pathways your organization is navigating.

Questions to Ask:

- What challenges are teams currently facing?
- Where are people struggling most: igniting change, building trust, adapting, or staying committed?
- What strengths can we leverage?

Tools to Use:

- Surveys to gauge team readiness and mindset.
- SWOT (Strengths, Weaknesses, Opportunities, Threats) analysis for organizational insights.

2. Design a Tailored Workshop Series

Effective transformation journeys are not one-size-fits-all. Customize the workshops and tools from the Core Pathways to fit your organization's culture, goals, and challenges.

Consider:

- Team Dynamics: Adjust the focus to engage both skeptics (like Clara) and early adopters (like Kai).
- Business Context: Prioritize pathways most relevant to your immediate objectives. For example, a startup might focus on *Ignition* and *Creation*, while an established company could emphasize *Adaptation* and *Commitment*.
- Pacing: Determine the ideal sequence of workshops to build momentum without overwhelming participants.

3. Facilitate with Purpose

The success of your workshops depends on skilled facilitation that fosters trust, encourages collaboration, and drives engagement.

Best Practices:

- Use experienced facilitators who can navigate sensitive topics and maintain a positive tone.

- Encourage storytelling to connect participants' personal experiences to the transformation journey.
- Integrate real-world examples and organizational data to ground abstract concepts in tangible realities.

4. Build Ownership Across the Team

Transformation succeeds when everyone feels they have a stake in the process. Foster a sense of ownership by involving team members at every step.

Tactics:

- Role-Based Assignments: Match tasks to individual strengths to increase engagement and accountability.
- Cross-Functional Teams: Create diverse groups to ensure different perspectives are represented.
- Peer Leadership: Empower individuals to act as ambassadors for specific pathways, driving buy-in at all levels.

5. Measure Progress and Adjust

Transformation is iterative, not linear. Regularly track your progress, celebrate milestones, and adapt your approach based on feedback and outcomes.

Key Metrics to Monitor:

- Use surveys or polls to measure participation and sentiment.
- Observe shifts in collaboration, communication, and problem-solving.
- Track performance indicators tied to your transformation goals, such as productivity, innovation rates, or employee satisfaction.
- Treat feedback as a gift - every insight is an opportunity to refine and improve your approach.

6. Sustain the Journey

Transformation doesn't end with a workshop series; it's a continuous cycle of growth and improvement.

- **Embedding the Pathways:**
    - o Incorporate the Core Pathways language and principles into organizational practices, such as onboarding, performance reviews, and team meetings.
    - o Use symbols, visuals, or storytelling (like the *Horizon One* narrative) to reinforce the shared journey.
- **Sustaining Momentum:**

o Schedule periodic reflections to revisit progress and recalibrate strategies.

o Empower teams to take ownership of new challenges, applying the pathways independently.

By weaving the Core Pathways Model into your organization's culture, you're not just implementing change, you're building a foundation for continuous evolution and success. Transformation becomes more than a goal; it becomes a mindset.

Conclusion: Charting Your Path Forward

The journey of the *Horizon One* reflects the complexities, challenges, and triumphs of navigating change. It is both a story and a mirror, offering inspiration through its narrative and guidance through its framework. With the Core Pathways Model, we've explored the essential steps to transformation - embracing ignition, fostering connection, adapting with resilience, and creating a future rooted in capability, commitment, and success.

Yet, like any journey of change, this story is not an end, it's a beginning.

Change is relentless, a constant force shaping the way we live, work, and grow. As leaders, collaborators, and creators, our task is not to resist it but to navigate it with intention, to transform its uncertainty into opportunity, and to inspire those around us to rise to the challenge.

The tools and insights shared in these pages are not just concepts; they are blueprints for action, designed to empower you and your teams to chart a course through the unknown with courage and confidence.

Every organization is its own *Horizon One,* navigating uncharted space toward innovation, growth, and purpose.

Whether you are a visionary champion like Theo, a thoughtful skeptic like Clara, or a dynamic early adopter like Kai, your role in this journey is critical. The Core Pathways Model serves as a map, but the success of the journey depends on the collective strength and commitment of the team steering it forward.

So where do we go from here?

As you reflect on the Core Pathways Model and the story of the *Horizon One,* let this be a call to action, a moment to ignite your next transformation:

- Reflect on where you and your organization are in your journey. What challenges do you face, and where do you see opportunities to grow?

- Connect with your team to align on shared goals and build the trust and cohesion needed for success.

- Act with purpose, using the Core Pathways as your guide to tackle each stage of transformation with clarity and resolve.

Transformation is not just about reaching a destination; it's about the process of getting there.

It's in the choices we make, the people we engage, and the challenges we overcome.

Each step forward builds the momentum for what's next, creating not only solutions but a culture of adaptability and progress.

Are you ready to chart your path forward? The stars are waiting.

About the Author

Chris Broyles is a creative, communications, and marketing strategist dedicated to helping organizations navigate transformation and change. As the founder of **B Degree Publishing**, an imprint of **B Degree Creative & Communications**, he specializes in strategic storytelling, persuasive media, and high-impact communications that drive change and engagement.

Chris has worked with Fortune 500 companies, professional services firms, law firms, and leaders across industries to amplify their stories' impact whether in the courtroom, the conference room, or the boardroom. His expertise lies in crafting compelling narratives and visuals that bring clarity to complexity, whether through executive messaging, video storytelling, or branded content.

Passionate about the intersection of creativity, psychology, and technology, Chris explores how AI is reshaping business, collaboration, and the future of work. When he's not writing or consulting, Chris enjoys exploring Chicago's food scene, developing creative side projects, and spending time with his family.

www.ingramcontent.com/pod-product-compliance
Lightning Source LLC
Chambersburg PA
CBHW071339210326
41597CB00015B/1503